Phil

Science

A Beginner's Guide

ONEWORLD BEGINNER'S GUIDES combine an original, inventive, and engaging approach with expert analysis on subjects ranging from art and history to religion and politics, and everything in between. Innovative and affordable, books in the series are perfect for anyone curious about the way the world works and the big ideas of our time.

Philosophy of Science
A Beginner's Guide

Geoffrey Gorham

ONEWORLD

OXFORD

A Oneworld Paperback Original

Published by Oneworld Publications 2009

Copyright © Geoffrey Gorham 2009

The moral right of Geoffrey Gorham to be identified as the
Author of this work has been asserted by him in accordance with the
Copyright, Designs and Patents Act 1988

ISBN 978–1–85168–684–1

Typeset by Jayvee, Trivandrum, India
Cover design by www.fatfacedesign.com
Printed and bound in Great Britain by TJ International, Padstow

Oneworld Publications
185 Banbury Road
Oxford OX2 7AR
England
www.oneworld-publications.com

Learn more about Oneworld. Join our mailing list to
find out about our latest titles and special offers at:

www.oneworld-publications.com

Mixed Sources
Product group from well-managed
forests and other controlled sources
www.fsc.org Cert no. SGS-COC-2482
© 1996 Forest Stewardship Council

Contents

Illustrations

Introduction: What is philosophy of science?

While science is a relatively recent product of human culture, the drive to understand our natural surroundings seems to be a deep and distinctive feature of human nature. As Aristotle – perhaps the first great scientist (and philosopher of science) – observed "All men by nature desire to know." This same curiosity that encourages science gives rise to philosophical reflection about science. Thus it has always seemed remarkable to scientists that humans have not only the desire but also the capacity to understand nature – and that nature has the capacity to be understood. In this vein, Albert Einstein declared paradoxically that "the eternal mystery of the world is its comprehensibility." Solving this mystery is one of the fundamental and abiding problems of philosophy of science. Perhaps, as philosophers even before Aristotle speculated, nature has an inherent language or *logos* that somehow suits it for human comprehension. Or perhaps scientific theories merely project human categories onto an otherwise indifferent and inscrutable world. Is the world discovered or constructed by science? As we will see in chapters four and five below, this ancient question is still hotly debated in philosophy of science today.

But a word about *philosophy*. Since the label is derived from the Greek words for love (*philia*) and knowledge or wisdom (*sophia*) a philosopher is literally just a lover of knowledge. But this doesn't capture the precise nature of philosophical inquiry. Presumably, non-philosophers such as doctors and lawyers (maybe even some politicians) love knowledge too. But whereas these fields are directed at knowledge or expertise in specific subjects, philosophy ranges very broadly over every area of human concern. With the possible exception of logic, philosophy doesn't really possess an established body of knowledge like math or history. Indeed, philosophers rarely agree among themselves about even the most basic issues in their field. What distinguishes philosophy from the other disciplines is its concern with *fundamental* problems and concepts at the core of any human activity or interest: the nature of knowledge, the structure of reality, the meaning and value of life, and so on. To be sure, philosophy puts forth its share of definite theories and claims (some possibly true; many quite outlandish). But these are properly *philosophical* theories and claims because they answer to the hope of understanding and clarifying the really basic issues. Put briefly, philosophy is the attempt to get to the bottom of things.

Accordingly, philosophy of science is the attempt to answer fundamental questions about science. Is scientific knowledge different from other sorts of knowledge? Is science getting closer to the absolute truth? Is science influenced by politics and gender? How are the various sciences related to one another? In addition, there are philosophical questions that arise within particular sciences, like psychology (could a machine think?), physics (is the world deterministic?) and biology (does evolution have a built-in tendency to complexity?). And in this book we will frequently have occasion to touch on these field-specific questions. But our main concern will be the "big questions" about the nature of science itself.

Although the philosophy of science only emerged as a professional sub-discipline of academic philosophy – with its own journals, courses, associations, etc. – in the twentieth century, it is really as old as philosophy itself (i.e. very old indeed). This is because philosophers, as lovers of knowledge, have always been impressed with the power and depth of scientific knowledge. Many of the greatest philosophers in the Western tradition were (at least part-time) philosophers of science: Aristotle, Descartes, Hume, Kant, Mill, and Russell, to name a few. In fact, the very distinction between philosophy and science is relatively new. Before the nineteenth century, there was simply philosophy or "natural philosophy." Isaac Newton's masterwork in mechanics is titled *The Mathematical Principles of Natural Philosophy*. And although philosophy may not come up routinely in the daily practice of science, problems at the cutting edge of science will often turn on deep philosophical issues about time and space, causality, and experience. So it's not surprising that some of the greatest scientists have been profoundly philosophical in their orientation. Along with Newton, this can also be said of Galileo, Darwin, Niels Bohr, Albert Einstein, Stephen Jay Gould, and Stephen Hawking. And as we shall see in a moment the very earliest scientists were simply philosophers who took a special interest in the natural world.

Philosophy, especially as it has been practiced lately, is a highly abstract enterprise. For the most part, this is as it should be, since philosophy is not concerned with this or that particular state of affairs but with very general concepts and problems. As the prominent twentieth-century philosopher Wilfrid Sellars put it, the aim of philosophy is "to understand how *things* in the broadest possible sense of the term *hang together* in the broadest possible sense of the term." This is true of the philosophy of science. Thus, the central chapters of this book, two through four, are concerned with three basic questions about how science hangs together in a very general sense. What is the essence of science? What is the method of science? What are the aims of science?

But the philosopher must not forget that science is also a concrete product of human culture, with a definite history and an immense impact on human (and non-human) welfare. So this book opens with an account of the origin of science and its gradual separation from religion and philosophy. And it ends with a consideration of the relation between science and human values. Chapter five considers to what extent science is, or should be, permeated by social and political forces, while chapter six asks what the likely future of science bodes for humanity. I have my own opinions on all of these issues, which will no doubt be evident from time to time in the course of the book. My goal, however, is not to convince you of any particular view but only show you that science, like nature itself, presents a deep source of philosophical wonder.

1

The origins of science

The most basic question one can ask about science is simply, what is it? One obvious way of identifying the nature of a thing is by attempting to define it. A good definition will tell us what is both adequate and essential for something to be the thing in question. For example, the definition of "collier" will tell us that all and only coal miners qualify. So we might explain what science is by identifying what is sufficient and necessary for something to count as genuinely scientific, i.e. to mark out exactly what does and does not fall within the boundaries of science. Analogously, if someone wanted to understand what Canada is, I could simply explain to them where its borders lie: Canada is the sum of territory lying inside these boundaries and no territories beyond.

As we will see in chapter two, just as national boundaries are often disputed and unclear, it is surprisingly difficult to arrive at a precise definition of science. Fortunately, there is another useful way to learn about the nature of something besides its definition, and that is by studying its history. We can learn about Canada by asking questions like: how did there come to be a territory that we now designate by the name of "Canada" and how have its contours and boundaries been shaped by older and neighboring countries, by economic and social forces, civil and foreign wars, and so on?

In this opening chapter we adopt the historical approach, learning as much as we can about the nature of science by exploring its origins and early development. Since our present aim is to identify the distinguishing features of science, rather than its internal evolution and refinement, our focus will be on the emergence of science from its roots in the religion and philosophy of ancient Greece through its flowering during the Scientific Revolution of seventeenth-century Europe. Subsequent chapters will consider more recent developments in modern science, once we have secured a basic familiarity with the birth and growth of this remarkable creature of human wonder.

Ancient beginnings

How did our world come to be, and why does it have the structure it has? These are questions of *cosmology*, the oldest of sciences. The earliest cosmologies took it for granted that the natural world was made in the way garments, dwellings, and tools are – by the design and efforts of intelligent beings. Since the task of fashioning the entire world was obviously an enormous one, the power of these hypothetical beings, or gods, made them a worthy object of fear and worship. Their efforts, like human affairs, were sometimes collaborative but often antagonistic. Thus, according to an ancient Babylonian creation myth, the supreme god Marduk made the earth and sky by splitting the body of his rival Ti'amat in two. The Sumerians had their 'Eridu Genesis,' and the Egyptians had Nun, who produced the world out of the limitless ocean while efficiently delegating some of the lesser tasks to subordinates.

In ancient cultures such anthropomorphic deities were invoked to account not only for the origin of the natural world but also for its changes and cycles over time. In order to chronicle divine activities and signs, and mark the religious festivals associated with agricultural cycles, the ancient Babylonians, Egyptians, and Syrians all produced highly detailed charts of the heavenly motions. Early astronomy was facilitated in this mapping of the stars by remarkable advances in arithmetic, geometry, and even algebraic formalism. So already in the pre-scientific world mathematics was becoming an indispensable tool for comprehending nature. But although the early astronomers increasingly relied on mathematics to *describe* the natural world, they continued to *explain* natural processes in supernatural and mythical terms.

The Greeks had their own gods, of course (notably Zeus and Apollo) and myth-makers (notably Homer and Hesiod). But a radically different approach to cosmological explanation that made little or no use of the traditional gods took hold in Greece in the sixth century BCE. A group of philosophers from the settlement Miletus on what is now the western coast of Turkey developed ambitious models of the universe that relied primarily on natural forces and entities rather than superhuman beings or gods. Each held that all natural phenomena are manifestations of a single underlying substance. Thales, for example, said the substance was water, perhaps because it was observed to transform from solid to liquid to gas. Earthquakes, he suggested, are from disturbances in the seas, and vision is from reflections in the aqueous material in our eyeballs. Another philosopher, Anaxagoras, preferred a less reductionist, somewhat 'chemical' conception of reality, account-ing for all substances as precise mixtures of the basic elements of earth, air, fire, and water. He also posited an underlying force, *nous*, which was not exactly a god but rather a ruling principle or aim for things. Another school employed the guiding assumption, which would re-emerge in the seventeenth century, that the world was composed only of tiny indivisible atoms swerving and crashing in a limitless void. Despite the differences among their models, all the early Greek cosmologists shared the aim of accounting for the observable world in terms of only a few, purely natural principles. And this has remained an aim of cosmology, and of science generally, ever since.

The Greeks were also skilled in mathematics, especially geometry. The fundamentals of pure geometry were set down by Euclid in his *Elements,* which provided a model of deductive reasoning from self-evident axioms or postulates, step-by-step, to sometimes surprising results. There is a famous anecdote reported by John Aubrey about the seventeenth-century philosopher Thomas Hobbes that nicely illustrates the power of Euclid's method. Glancing at an open copy of the *Elements* on his friend's desk, Hobbes read a surprising theorem displayed there. Declaring "that is impossible!" Hobbes doggedly traced the proofs backwards, theorem by theorem, all the way to the postulates on the opening pages, and was finally convinced. "And this," says Aubrey, "made him in love with geometry." For two thousand years, Euclid's theorems were thought to capture the only possible consistent geometry, until several non-Euclidean geometries were discovered in the nineteenth century.

In astronomy, geometry was applied with immense precision very early on by Eudoxus and Ptolemy. The latter's geocentric or "earth-centered" model of the planetary system, which combined common sense with empirical accuracy, guided astronomical inquiry through the sixteenth century and is even used still for convenience in some methods of marine navigation. Geometry and arithmetic were pursued with religious zeal by the followers of the philosopher Pythagoras (for whom the famous theorem about triangles is named). The Pythagoreans even postulated that the natural world is in some sense *made* of numbers and imagined that the planetary orbits produced music like the harmonized strings of a lyre. This Pythagorean faith that nature is fundamentally mathematical and comprehensible persists in major figures of the Scientific Revolution like Galileo and Newton, and in fundamental theories of modern physics like string theory, as we will later see.

But the most famous philosophical citizen of classical Athens came to reject the generally scientific orientation of early Greek thought. Socrates was less concerned with the structure of the universe than with the nature of virtue and justice. Indeed, his dogged pursuit of these ideals irritated the city fathers and culminated in his trial and execution for "corrupting the youth." One such youth, Socrates' great student Plato, recounted these dramatic events in a series of brilliant and moving dialogues usually collected together as *The Trial and Death of Socrates*. In the dialogue *Phaedo,* Socrates explains his disenchantment with the naturalistic approach of earlier philosophers: "I was wonderfully keen on the wisdom which they call 'Natural Science'. For I thought it splendid to know the causes of everything, why it comes to be, why it perishes and why it exists." But Socrates found that science could only explain how things seem to come together and separate, not why they have the natures they do. For example, why something is a "unit" or large or beautiful cannot be explained in terms of the physical or chemical composition of its parts. So Socrates concludes, "I do not anymore persuade myself that I know why a unit or anything else comes to be or perishes or even exists by the old method of inquiry, and I no longer accept it."

The disenchantment with science's pretensions to explain the fundamental nature of things is really as much Plato's as Socrates'. Although Plato devoted considerable attention to cosmology – positing a primordial craftsman or *demiurge* which brought order out of chaos – his own metaphysical viewpoint devalued scientific knowledge. In fact, Plato considered the familiar trees and rivers that we perceive by the senses at best poor imitations of the ideal "forms" of TREE and RIVER. Unfortunately, we will only know the forms adequately when death releases us from the "prison" of our body. Until that happy day purely intellectual modes of inquiry (philosophy and mathematics) are the closest we can get. (Socrates characterizes philosophy as "practice for death" in the *Phaedo* – at the climax of which Socrates himself expires – acknowledging wryly that many would say philosophers are already effectively dead.) In

Plato's famous allegory, most of us are like the cave-dwellers fixated on the shadows flickering on the wall, oblivious to their eternal and perfect source. In a possible dig at the scientists of his time, Plato remarks on the absurdity of honoring those in the cave "who are sharpest at identifying the shadows as they pass by and remembering which come earlier, which later, and which simultaneously."

It has been said that all of philosophy after the Greeks is a "mere footnote" to Plato. It may be said with equal justice that all of subsequent natural science can be traced to the inspiration of Plato's student Aristotle. Aristotle's surviving works seem to be based on lecture notes rather than formal treatises and so they can seem scattered and opaque. Nevertheless, they display a scientific intellect of unparalleled breadth and insight. A famous Renaissance painting by Raphael called the *School of Athens* beautifully encapsulates the divergent attitudes of Plato and Aristotle, teacher and student. As the grey-bearded Plato gestures reverently to the heavens, the realm of the forms, the young and virile Aristotle, at his teacher's side but also a modest step ahead, calls our attention to the immediate surroundings, the realm of the senses. In Aristotle's view, all knowledge comes from experience and careful study of the natural world is a source of both wisdom and delight. The son of a physician, he was especially fascinated by living things: "We should venture on to the study of every kind of animal without distaste," he remarks in the *Parts of Animals*, "for each and all will reveal to us something natural and something beautiful."

The point of detailed observation is not merely delight however; the true scientist aims to identify the real *causes* of natural phenomena. Aristotle held that there are four causes or reasons for any phenomenon: material, efficient, formal, and final. Consider animal reproduction and generation, which Aristotle studied very closely. The *material cause* is the "stuff" involved in the process; in most cases of generation this is the egg from the female. The *efficient cause* is the immediate source or "trigger" of motion or change; the sperm from the male according to Aristotle. The *formal cause* is what makes something the kind of thing it is. In the case of animals, the defining features of its species constitute the formal cause. For example, rationality and two-leggedness are parts of the formal cause of human beings. Although formal causes serve a function analogous to Plato's forms in accounting for the natures of things, Aristotle denies that formal causes exist on their own, "separated" from material things.

Finally, the *final cause* of a natural process is its end or purpose; in animal generation this is the adult organism. Aristotle did not believe in an "intelligent designer" of the universe but he did posit ends or purposes throughout nature. The various parts of animals have their own final causes (the heart's purpose is to pump blood, the eyes to see, etc.) as does reproduction as a whole (immortality of a sort). Purely physical processes have ends too: the planets aim to achieve perfection in their perpetual circular motions, falling bodies aim to reach their place of "natural rest" at the center of the earth, and fire strives upwards. The motion of the world as a whole depends on a divine "unmoved mover," which acts as a final cause – the object of desire to which the world is drawn. The notion that un-designed, unconscious processes have aims and goals may seem bizarre to the modern mind, but Aristotle thought purpose was essential to explain natural processes. As we shall see, the abandonment of final causality is a major turning point in the emergence of modern science.

Aristotle was the first philosopher to expound in detail on the scientific method. Scientific reasoning, in his view, involves crucially the use of arguments or syllogisms. These combine universal premises (*all mammals nurse their young*) with particular or less universal premises (*dogs are mammals*) to infer some fact (*dogs nurse their young*). Note the logical force of such arguments: if the premises are true then the conclusion must be true also. In the case of scientific demonstrations, Aristotle says the premises will be necessary though not self-evident. Since the premises are necessary, the conclusion will be necessary too and we can claim secure knowledge of the facts deduced. But this raises an obvious question: how can we know the premises of a scientific demonstration to be necessary? Aristotle does not think this is known by generalizing from observed instances, since we can never be certain we have encountered all relevant instances. In a discussion that is rather obscure even for Aristotle, he suggests a more direct, intuitive way of apprehending the basic principles of science: "since except intuition nothing can be truer than scientific knowledge, it will be intuition that apprehends the primary premises." Here Aristotle is perhaps showing the influence of Plato, contrary to his own more empirical, "down to earth" conception of knowledge. Still, his question how science can be both certain and empirical becomes a perennial one for the philosophy of science, and we will return to it in chapter three.

Despite his immense influence on later science, there are two important features of modern science that are quite alien to Aristotle: experimentation and mathematical law. In his studies of generation, he meticulously observes and catalogs changes in the chicken embryo, but never subjects the egg to different environments to see how this affects its development. Part of the reason for this reluctance to experiment is that for Aristotle science is exclusively concerned with *natural* changes and processes rather than artificial constructions. So if we excessively manipulate natural conditions by designing elaborate experiments we are not really doing science but something more like art or craft. As for the absence of mathematical laws, Aristotle does offer laws of motion but they are expressed strictly in terms of proportions of unquantified qualities rather than numerically. He seems to have considered mathematics of only limited relevance to science, at least outside the realm of astronomy where motions are highly regular and simple. Mathematics is concerned with idealizations or abstractions but natural change is complicated and subject to numerous influences of various kinds. Combustion, for example, is a complex change involving earth, air, and fire (and opposed by water). How could number tell us anything about it? Thus, Aristotle dismisses the Pythagoreans: "They have said nothing whatever about fire or earth or other bodies of this sort, I suppose because they have nothing to say which applies peculiarly to perceptible things."

Aristotle worked in every area of science, including cosmology, physics, anatomy, and psychology; but he was especially

BREAKING THROUGH TO THE OTHER SIDE (OF THE UNIVERSE)

Aristotle, like most ancient cosmologists, held that the universe is a finite sphere bounded at the outermost edge by a fifth element (quintessence) different in kind from the other four. Since he rejected the notion of void or empty space, Aristotle thought it was meaningless to speculate about the place beyond the outermost sphere – there simply is no "there" there. Defenders of an infinite, void space, from Aristotle's time through the seventeenth century, have frequently relied on a certain thought experiment that is customarily attributed to the Greek mathematician Archytas. Suppose a swordsman is positioned at the edge of the quintessence. Could he extend the sword outward? If he can, then there is space beyond the sphere after all. Moreover, since it seems this could be repeated indefinitely the universe must be infinite. If he can't extend the sword then there must be some solid barrier preventing this. But this barrier must have an outer edge beyond the tip of the sword. So apply the same thought experiment at this more outward edge, and so on. This and similar thought experiments were debated throughout the Middle Ages and eventually invoked by Locke and Newton in support of the modern notion of infinite, absolute space (and time).

attracted to biology. Yet even though his father was a physician, his own medical investigations were somewhat limited. But ancient Greece had other great medical scientists. The development of ancient medicine followed a similar path to that we have seen in cosmology, gradually moving away from supernatural to natural explanations. The medical arts of Egypt and Mesopotamia were a mix of physiology, demonology, and magic, with diagnosis and treatment alike relying on supernatural hypotheses. But the Greek medical genius Hippocrates, source of the still-sacred physician's oath, looked strictly for internal, physical causes based on meticulous examination of physical symptoms. He held that illness was normally caused by imbalances in body chemistry and best healed by encouraging the body's own immune systems rather than by surgery or invasive procedures. This holistic, systematic approach to medicine is epitomized in the "four humors" model of disease, developed by Galen and others, which dominated medical science for a millennium. In this view, just as there are four basic elements in the non-biological realm (earth, air, fire, and water) there are four chemical elements in delicate balance in the human body: black bile, yellow bile, blood, and phlegm. (Traces of humoristic psychology remain in our language if not our

ZENO'S PARADOXES OF MOTION

In denying that the world of the forms is subject to change, Plato was deeply influenced by the philosopher Parmenides. Reflecting deeply on the difficulty of talking or conceiving of what is "not" – even thoughts about fictional beings are not about just nothing – Parmenides was driven to the radical conclusion that there is only one thing, which he dubbed "It-is." After all, he reasoned, if there were two things, each would be "not" the other and you can't really conceive what is not. Furthermore, "It-is" is unchanging since otherwise it would be first some way and then "not" that way.

ZENO'S PARADOXES OF MOTION (*cont.*)

Parmenides' reasoning is obviously hard to swallow. But his brilliant student Zeno constructed a series of famous arguments in defense of his teacher aimed at showing that the belief in multiplicity and change leads to absurdity.

Several arguments were directed against the possibility of motion, which Parmenides rejected as a sort of change. One of these famous "paradoxes of motion" centered on a race between Achilles (fleetest of the Homeric heroes) and a tortoise. Suppose the race is 20 meters long and Achilles sportingly offers the tortoise a 5 meter head start.

$$T1 _____ T2 \_\_\_ T3 \_\_ T4 \ldots _____$$
$$A1 _____ A2 _____ A3 \_\_\_ A4 \ldots _____$$

However fast Achilles runs, in the time he takes to reach 5 meters, the tortoise has moved ahead some distance, let's suppose 2.5 meters. In the time required to reach that position, the tortoise has again moved 1.25 meters. And so on *ad infinitum*. So Achilles can never catch up, Zeno reasoned, since he would need to cover an infinite distance in a finite time. This is the absurdity: a faster runner can never overtake a slower. The paradox depends on a mathematical error that will seem obvious to a reader familiar with modern mathematics. $5 + 2.5 + 1.25 + \ldots$ does not sum to infinity but rather to 10 (the "limit" as understood by modern calculus). So Achilles will overtake the tortoise at 10 meters. The fallacy was perceived, albeit somewhat dimly, by Aristotle, who observed that Achilles has as much time as he needs to overtake the tortoise since time is infinitely divisible just as much as space: there is no lack of time in which to cover all the infinite parts of space.

But there is another way of framing the paradox, not so easily resolved. Suppose Achilles carries a baton in his right hand which he switches to his left hand at 5 meters, back to his right at 7.5, and so on. Which hand holds the baton at the instant he finally reaches the tortoise at 10 meters? There seems to be no answer to this question. But surely there must be! This and similar puzzles deriving from Zeno remain at the heart of ongoing inquiry into the nature of space, time and infinity.

medical practice: "bilious;" "phlegmatic," etc.) Although this approach sometimes involved practices that may seem crude from the point of view of modern medicine (e.g. bloodletting), medical science is still largely concerned with imbalances among various elements (hormones, antibodies, neurotransmitters).

The Middle Ages and Renaissance

In the early centuries of the first millennium, the essential elements of Greek philosophy and science (not to mention art and literature) were adopted and transformed by the Roman civilization, while in Greece itself the philosophies of both Plato and Aristotle were developed and refined. But with the decline of the Roman Empire, and the rise in political power of the main religious orthodoxies of Judaism, Christianity, and Islam, Greek learning was increasingly neglected or disparaged. "What has Athens to do with Jerusalem," asked the early Christian leader Tertullian, "or the Academy with the Church?" The influence of Greek thought upon figures like St. Augustine and Boethius is significant but primarily theological in orientation. Science becomes at best the "handmaid of theology."

From the fifth century through the end of the first millennium, independent scientific inquiry remained more or less dormant in Europe. But in the Islamic world, around the ninth century and especially in Syria and Baghdad, Aristotle's texts began to be widely translated and disseminated. This led, as part of the broader "Islamic Golden Age," to a flowering of philosophical activity, with Islamic philosophers Al-Farabi, Avicenna and Averroes, along with Jewish thinkers like Maimonides, debating the theological implications of Aristotelian philosophy and cosmology. Science and mathematics also flourished around the turn of the millennium with astronomy and medicine producing serious challenges to Ptolemaic geocentrism and to

Galenic humorism. However, the Golden Age's comparatively liberal attitudes concerning the relation between science and religion eventually gave way to theological orthodoxy, and under the added pressures of the crusades and Mongol invasions, Islamic science declined rapidly in the twelfth and thirteenth centuries.

But just as learning waned in the Middle East, a revival was underway in Europe. Aristotelian texts preserved through the Golden Age were translated into Latin and widely studied at European schools and monasteries that had previously been dominated by Platonically oriented Christian theology. Whereas in Islamic thought natural philosophy or science was generally held at "arms-length" from theology, the empiricist and natural-istic inclinations of Aristotle's thought were assimilated into Christian theology. St. Thomas Aquinas (who frequently and reverently cites Aristotle as simply "The Philosopher") regarded science and religion – or more broadly reason and faith – as complementary paths to knowledge of God and creation. Since man is made in the image of God it is appropriate to use our intellect to understand the world. And it befits the workmanship of God to invest the creation with real powers and natures, which it is the aim of science to discern. Philosophers more directly involved in the sciences, like Robert Grosseteste and Roger Bacon, advocated experimental science both for its own sake and as a suitable companion to religion.

This new intellectual atmosphere, as well the dramatic growth in urban centers and trade, encouraged the establishment of major universities at Oxford, Paris, and Bologna, where trad-itional theology was pursued alongside increasingly technical scientific programs. In the fourteenth century a group now known as the "Oxford Calculators" introduced into physics the fundamental distinction between kinematics (free motions) and dynamics (motions under the influence of forces). This allowed for the first time since Aristotle purely mathematical analysis of

CONDEMNATIONS OF ARISTOTLE

Aristotelian philosophy was not by any means embraced universally during the Middle Ages. Occasionally, supporters of "pagan" thought like Roger Bacon were persecuted and imprisoned. And official condemnations of specific Greek-influenced doctrines and texts were routinely issued by the Catholic Church. For example, in 1277, Bishop Tempier of Paris formally condemned 219 theological and scientific "propositions" many of which he associated with "radical Aristotelianism." Of particular interest are his condemnations of the following propositions:

- That God could not move the heavens with rectilinear motion because in that case a void would remain.
- That if heaven stood still fire would not burn flax because in that case time would not exist.

These propositions derive from Aristotle's position that a void without matter and time without motion are both impossible. Tempier rejected these doctrines for imposing undue limitations on God's power: why couldn't God, if he wanted, create a region of void space or a stretch of "empty time?"

While such condemnations might seem an illegitimate intrusion of theology into the domain of science, ironically they may have played a role in the birth of modern science, according to eminent historians like Pierre Duhem. For they encouraged scientists and philosophers to explore the physical and cosmological implications of God's omnipotence. For example, if God can move the entire universe, and also stop and resume its motion as he pleases, it seems there must be empty space and time in which he could realize this power. Likewise, if contrary to Aristotle the heavenly spheres can be moved by God's will alone, without the influence of other forces, then there is no reason why they should not continue in this motion forever. The condemnations thus helped set the groundwork for the modern notions of inertia and absolute space and time, which had been precluded by Aristotelian philosophy.

motion independent of the question of its "causes," and this in turn led to close approximations of the law of falling bodies. Jean Buridan developed impetus theory, forerunner to the modern notion of inertia, to solve a problem that dogged Aristotelian physics: why does a projectile continue to move upwards after release, away from its position of natural rest? And Nicole Oresme provided the first systematic arguments for the earth's motion.

As part of the broader "rebirth" (renaissance) of learning in Europe, science blossomed and diversified in the fifteenth century. Plato replaced Aristotle as the major source of philosophical inspiration, though Platonism was often blended with magical, Christian, Jewish (cabbalistic), and Middle Eastern elements. This eclecticism was accompanied by a general distrust of intellectual authority in favor of individual reason and direct observation. The spirit of innovation was enabled in part by the growing use of scientific applications in navigational and military technologies – scientists could look for financial support beyond the church and universities. This adventuresome intellectual atmosphere brought major discoveries in every field of science (not to mention art and literature) but also generated conflict as traditional authorities attempted to shore up power. The most famous illustration is the revolution in astronomy, which removed the earth from the center of the universe and set science on a path of independence unknown since the classical period.

The Copernican Revolution

Ptolemy's geocentric model of the heavens enjoyed the combined support of common sense, Aristotle, and seemingly the Bible itself, where, for example, God is credited with making the sun "stand still" for a day (Joshua 10:12–13). But although it had served for hundreds of years as a highly accurate predictor of planetary motions, as observational techniques improved Ptolemaic astronomers were forced to make numerous ad hoc (after the fact) adjustments to preserve the geocentric model. For example, from the Earth's perspective, several planets display a "retrograde" motion, zig-zagging and backtracking across the night sky rather than following continuous circular paths. To explain this, astronomers imposed "epicycles" on the orbits, so that the planets traced circles upon circles and seemed to "loop-the-loop." As anomalous observations accumulated, more and more epicycles were introduced, along with various other mathematical devices. The Polish astronomer Copernicus compared the immensely complicated model that resulted from all these adjustments to a single human portrait drawn from many beautiful but divergent models – "the result would be a monster rather than a man."

Copernicus quietly introduced his own sun-centered model in 1514 in a brief, handwritten pamphlet. The argument was worked out in much more detail and finally published in 1543 as a large technical treatise called *Revolution of the Heavenly Spheres* (it is said that Copernicus died on the very day he finally saw a printed copy). One can only speculate how Copernicus

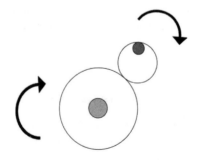

Figure 1 A typical orbital epicycle

reacted to the following apologetic line from a preface inserted without his consent by his editor, the German theologian Andreas Osiander: "hypotheses need not be true nor even probable. On the contrary, if they provide a calculus consistent with the observations, that alone is enough." Osiander's suggestion that the aim of scientific theories is mere consistency with observation, rather than truth, though certainly not Copernicus' position, is nevertheless a useful stance to adopt when those theories conflict with orthodoxy. As will see, Galileo's own refusal to assume such a stance led to his condemnation 100 years later.

If consistency with observation was the sole aim of astronomy, the Copernican model stood little chance of overthrowing the Ptolemaic model that had been continuously "retro-fitted" to the data. But the situation was changed dramatically by the work of the German astronomer Johannes Kepler. Relying on the meticulous new observations obtained by the aristocratic Dane Tycho Brahe from his observatory on the island of Hven, Kepler proposed that the planetary orbits are slightly elliptical rather than strictly circular. Once freed from the Aristotelian prejudice for circularity, the Copernican model was now at least as accurate as the Ptolemaic. Furthermore, Kepler discovered that the elliptical orbits of the planets are related to their speed in the following way: the planets "sweep out" equal areas in equal times. This surprising and elegant law is just the sort of relationship Kepler himself expected to find: he held a Pythagorean conception of nature and even hoped someday to detect the "music of the spheres."

But the final triumph of the Copernican system waited on its most brilliant champion, Galileo. Born in Pisa in 1564, Galileo was the son of a musician and theorist Vincenzo Galilei (who was something of an iconoclast in his own field). He began his studies at the University of Pisa in medicine – though there is no evidence he dropped balls from the famous leaning tower there

– but soon switched to mathematics and science and gained a minor reputation from lectures discussing the location and dimensions of hell according to Dante's *Inferno*. At his next post at the famous University of Padua, Galileo perfected rudimentary telescopic technology for the purposes of astronomy, and wrote a treatise called the *Starry Messenger*. Armed with his telescopic observations, he began to publicly defend the Copernican system. He observed distinct phases of Venus (which are hard to reconcile with a stationary earth), the moons of Jupiter (which he named the "Medicean stars" to curry favor from the local nobility), and mountains on our own moon (which put the lie to the longstanding Aristotelian doctrine that the celestial and terrestrial realms have entirely different natures).

Unlike Copernicus, Galileo was a skilled expositor with a solid scientific reputation and mounting fame. The Catholic Church, feeling the pressures of the Reformation, was compelled to act. In 1616, Galileo's works were officially banned and he was personally ordered by Cardinal Bellarmine of the Italian Inquisition not to "hold or defend or teach" the Copernican system. Galileo did not take this lightly: in 1600, the unorthodox natural philosopher Giordano Bruno was burned at the stake by the Inquisition and Bellarmine was one of the judges in that trial. In the years following Bellarmine's command, Galileo devoted his immense talents to less politically sensitive scientific problems, including pure mathematics, hydrostatics, the nature of motion, and the structure of matter. But he eventually returned to the Copernican controversy, encouraged by the election of an old ally to become Pope Urban VII. Probably hoping for a compromise, the new Pope urged Galileo to adopt a philosophical stance like the one Osiander had falsely attributed to Copernicus. Galileo could discuss the Copernican system, but he should not claim it proven from the telescopic observations, since "God can conceivably have arranged things in an entirely different manner yet while bringing about the effects that we see." Bellarmine recommended the same stance: "that the Earth moves and the Sun stands still saves all the celestial appearances better than do eccentrics and epicycles is to speak with good sense and to run no risk whatever." The risk, as Galileo was soon to discover, is alleging the truth of the Copernican hypothesis.

GALILEO AND THOUGHT EXPERIMENTS

Galileo possessed gifts rarely combined in a single scientist: he was a keen observer, brilliant mathematician, ingenious experimentalist and technician, and a beautiful writer. He was also one of science's great devisers of "thought experiments." Without constructing an actual experiment, scientists sometimes reflect on what sort of phenomena should be expected in a highly abstract or idealized situation. Einstein once thought about riding on a light beam, for example, in order to draw out the implications of his idea that the speed of light is the same for moving and resting observers. The point of such thought experiments is usually to refine the theory rather than test it – to see what follows logically from the theory (or an alternative). But it might also lead to empirical tests, as happened with Einstein's thought experiment. It may reveal, for example, that the theory leads to predictions that are logically impossible. Galileo employed such a thought experiment against the prevailing assumption that bodies of different weights naturally fall at different rates. He imagined two cannonballs, one 10 lbs and one 1 lb, connected together by a firm rod. He then asked: will this object fall faster or slower than a single 10 lb cannon-ball? On the one hand, the received view says it will fall faster since it weighs more. On the other hand, the smaller ball should impose a slight "drag" on the larger ball and the whole object should therefore fall more slowly than the single 10 lb ball. There must be something wrong with the received view of free fall since we can see, without ever having to leave the comfort of our armchairs, that it leads to contradictory consequences.

Nevertheless, in 1632 Galileo published his *Two Chief World Systems*, a highly readable dialogue presenting an overwhelming barrage of empirical and conceptual arguments that finally sealed the fate of the Ptolemaic system. This clearly violated the 1616 command, and Galileo was duly called before the Inquisition in Rome, tried by Bellarmine, condemned, and ordered to recant his Copernican views (which he did: "I curse and detest said errors"). Although he was personally devastated by the condemnation, and his health suffered as a consequence, Galileo continued to work under house arrest at a villa near Florence (deriving some consolation from the proximity of his daughter, a nun) until his death in 1638.

The official issue in the Galileo trial was simply the violation of Bellarmine's order, but at stake was the overarching relationship between science and religion. In a widely circulated letter to the Grand Duchess Christina, Galileo openly argued that science should trump on questions about natural processes. Rather than deferring to the Bible about natural truths, he proposed reversing the order of command by making science a guide to biblical interpretation: "having arrived at any certainties in physics, we ought to utilize these as the most appropriate aids in the true exposition of the Bible." Also at stake was the nature of science itself. Rather than merely "saving the appearances" Galileo maintained that science should aim for "demonstrated truth" about the world. As we shall see in chapter four, this controversy continues to this day.

Galileo recanted in the end, but he was far from servile to his accusers, and in scientific controversies he could be extremely willful and even arrogant. Perhaps it's appropriate then that his middle finger has been preserved at the museum in Italy which houses his papers.

The Scientific Revolution

The Copernican Revolution removed the Earth from center stage to the third row of the natural universe. This encouraged the notion, entirely contrary to Aristotelian science, that the same forces and principles apply here as everywhere else. Galileo's other great accomplishment was uncovering the most general laws of motion for bodies near the earth and extending theses laws to all motions wherever they occur. Using a series of experiments involving pendulums, inclined planes, and projectiles, Galileo demonstrated that all bodies fall at the same rate of acceleration (in a vacuum) independently of their weight and composition. His approach to this problem involved two crucial departures from traditional or Aristotelian science. First, he was primarily concerned with arriving at precise mathematical descriptions of the phenomena rather than their explanation or cause. Although Galileo was sensitive to Aristotle's point that nature was far "messier" than the objects of pure mathematics, he never doubted that at bottom the book of nature is "written in the language of mathematics." Second, while he appreciated Aristotle's emphasis on the distinction between the natural and the artificial, he conceived of experimental constructions as essential to isolate and gauge the fundamental laws of physics.

Rather than simply eschew causal and metaphysical thinking, the great French philosopher René Descartes believed the Scientific Revolution could never succeed unless the doubtful Aristotelian framework was replaced with an entirely new conception of nature. Whereas Aristotle and his medieval followers had imbued nature with a multiplicity of natures, forces, and ends, Descartes insisted on an extremely austere and mathematical conception of the world as simply *res extensa* (extended being) in motion. Any phenomena not reducible to matter in motion, such as thoughts and volitions, he conceived as radically separate from the natural world and off-limits to science. This strictly "mechanical" view of nature was extremely fruitful in scientific terms, but also led to some surprising hypotheses on Descartes' part. For example, he was driven to regard animals as mindless machines (bête-machines). On the traditional view, humans, animals and plants each have their own kind of soul. But for Descartes having a soul is an all-or-nothing matter and he was averse, on religious grounds, to attributing them to animals: "it is more probable that worms, flies, caterpillars and other animals move like machines than that they all have immortal souls." Descartes also rejected the view of William Harvey (who discovered the circulation of blood) that the heart acted like a pump. Descartes preferred to think of the heartbeat as a perpetual fermentation and explosion rather accept Harvey's conception which seemed to give the heart a will of its own. But despite these excesses, mechanical philoso-phy became the guiding framework of the Scientific Revolution and was employed in the explanation of collisions, air pressure, gravity and chemical reactions. While Descartes thought the world was a "plenum" absolutely filled with matter, English scientists of the same period generally preferred a "corpuscular" version of mechanical philosophy which traces all phenomena to interactions among tiny particles. Robert Boyle used such an approach to discover the gas law that bears his name while Newton developed a corpuscular theory of light.

Another influential feature of Descartes' philosophy was its conception of the relation among theology, philosophy, and science, a highly controversial matter since the Middle Ages. In his *Principles of Philosophy,* he compared all knowledge to a tree "whose roots are metaphysics, whose trunk is physics, and whose branches, which grow from this trunk, are all of the other sciences." Descartes himself tried to derive the laws of nature, and hence all more specific phenomena, from certain allegedly self-evident propositions about God's nature and action. Rather than legislating the outcome of science in advance on the basis of bib-lical or Aristotelian authority, theology and metaphysics became integral to the scientific project. The philosopher John Locke rejected Descartes' project of deriving physics from metaphysics, but he accepted the revised model of philosophy as serving rather than commanding science. Thus, he writes in the preface to his own magnum opus, "Unlike the master-builders who work in the sciences, such as the great Huygenius and the incomparable Mr. Newton, it is ambition enough for philosophers to be employed as under-labourers in clearing the ground a little and removing some of the rubbish that lies in the way to knowledge."

The new science was also useful to religion. For the detailed understanding of nature was taken to provide evidence and insight about its creator, just as we learn of an author from reading her works. With nature serving as the "other book of God," science becomes a kind of "natural theology." As Francis Bacon put it, science is "after the word of God, at once the strongest remedy against superstition and the most proven food for faith." The search for divine purpose in Nature, sometimes referred to as "teleology," was highly contested during the Scientific Revolution. Although he grounded the laws of nature on God's immutable and continuous creation of matter and motion, Descartes considered it presumptuous in the extreme to speculate about God's purposes. The hyper-Cartesian philosopher Spinoza asserted that invoking God's will in science is taking refuge in the "sanctuary of ignorance," while Voltaire likened teleology to explaining the human nose as an ideal support for eyeglasses.

The final blow to teleology came in the nineteenth century with Darwin's theory of evolution by natural selection. But at its dawning modern science still involved a combination of empirical, philosophical and teleological thinking. A good illustration of this is the correspondence between Samuel Clarke, a supporter and friend of Newton, and the great German polymath Gottfried Leibniz about the existence of absolute time and space. One of Leibniz's main arguments against these Newtonian doctrines was that if there were absolute time and space, then God would have no reason to create the universe when he did in absolute time nor position it where he did in absolute space. But Leibniz declared it absurd that "God wills something without any sufficient reason for his will." As was typical during the Scientific Revolution, this controversy about absolute space and time turned on a complex mixture of factual, theological and methodological considerations.

The "incomparable Mr. Newton" mentioned by Locke above is of course Isaac Newton, the greatest genius of the Scientific Revolution. While the product of a modest rural background, Newton's mathematical gifts were evident from an early age. In order to arrive at the true science of natural motions, Newton realized he needed to invent a new mathematical technique for representing in a powerful and unified way rates of acceleration, instantaneous velocity, parabolic trajectories, and so on. This "method of fluxions," which we now call calculus, was invented by Newton in 1664, during the *annus mirabilis* (miraculous year) he spent quarantined at his rural cottage while the plague raged at Cambridge University. With this tool in hand, and building on the work of Kepler, Galileo, and his countryman Robert Hooke, Newton went on to publish the *Mathematical Principles of Natural Philosophy*, or simply the *Principia*, in 1687. Relying on three laws of motion, one law of "universal gravitation" and models of space and time as infinite and absolute containers, he was able to unify and explain the elliptical orbits of the planets, Galileo's law of falling bodies, lunar and solar eclipses, the daily motions of the tides, and the paths of cannonballs.

Like his compatriots Boyle and Locke, Newton generally adhered to a corpuscular and empiricist methodology and rejected appeal to intrinsic forces and powers in things (which he associated with scholasticism). Thus, he declared in the *Principia*: "I feign no hypotheses." However, early critics of the *Principia* could not help but notice that his system relied on a rather mysterious universal force acting at a distance without mechanical contact. The problem of gravity vexed Newton and has figured in many of the major advances in physics over the next 300 years, including the theory of relativity and the cutting-edge "string theory." The legitimacy of using hypothetical reasoning generally has been a major topic in modern philosophy of science, as we will see in chapter three.

In the years following the appearance of the *Principia*, Newton's laws were confirmed in numerous domains. Comets had been a source of puzzlement since the time of Aristotle. With their irregular yet apparently orbital paths they defied the traditional distinction between celestial and terrestrial motions. But Newton's laws respected no such dichotomy: "Nature is exceedingly simple and conformable to herself. Whatever reasoning holds for the greater motions holds for the lesser as well." In 1705, Newton's friend Edmond Halley used Newton's laws to prove that three previous comet sightings were actually a single comet. He then predicted that its fourth return would be in late 1758, much too late for him to observe personally. The comet was spotted on Christmas Day of that year and quickly named in Halley's honor. From the same laws and observations, astronomers could also "retrodict" earlier appearances of this comet and others, thus explaining observations recorded by Kepler, and many medieval and ancient scientists.

Newton pursued numerous unorthodox interests, such as alchemy and biblical criticism, with the same passion as mechanics. But his other lasting achievement was in the ancient science of optics. Combining careful experimentation and refined mathematical techniques, he explained the composition of white light by the colors of the spectrum and developed a corpuscular (as opposed to wave) theory of light. The achievements of Newton, as they became clear and widely known, were cause not only for scientific celebration but also national pride. The poet Alexander Pope captures the widespread enthusiasm for Newton, especially in his native land, with the following verse:

> Nature and Nature's laws lay hidden in night;
> God said Let Newton Be! And all was Light!

Newton's mechanics not only provided the basic theoretical framework for the next 200 years of physics (his laws are still used today in engineering problems where effects requiring Einstein's theory of relativity are negligible), but also set the mold for modern science as a whole. In the modern conception, science should provide laws that are universal. The laws should be given a precise mathematical formulation and allow detailed application to complex systems. Hypothetical entities and forces are permitted, but only if their effects can be directly observed and measured. The ultimate judge of a theory should be direct experience rather than philosophical, political or religious authority. Finally, modern science largely adopted the corpuscularian assumption that the features of large-scale objects and processes should be accounted for in terms of the motions and interactions of their smaller parts. In the chapters that follow, we will evaluate many important challenges to this image of science. Although modern science was 2000 years in the making, it has become the predominant mode of knowledge only in the last few centuries (and more recently than that in much of the world). So it is important to consider carefully and critically the nature of scientific knowledge, its relation to other ways of understanding, and its potential to guide humanity into the future. That is our pleasurable task in the following pages.

2
Defining science

In one of Plato's dialogues, the title character Meno poses a dilemma about the possibility of inquiry: "a man cannot search either for what he knows or for what he does not know. He cannot search for what he knows – since he knows it, there is no need to search – nor for what he does not know, for he does not know what to look for." Meno's dilemma neglects that normally when we inquire into the precise nature of something, we already have a rough idea what we're looking for. Our survey of the origins of modern science began with the distinction between science and myth, and along the way we refined our conceptions of the relations among science, mathematics, and technology. In this way, we have been able to draw some preliminary but important conclusions about the nature of modern science. For example we have found that although science has favored naturalistic, non-anthropomorphic explanations from the outset, the preference for experimental manipulation and mathematical laws is a more recent development. And we have seen that the role of hypotheses and teleological reasoning remained controversial even through the Scientific Revolution. Science came into its own in the seventeenth century, but it did not sever all connections to its origins in philosophy and religion.

But the historical approach has its limitations. Returning to our earlier analogy, serious disputes can arise about the exact boundaries between nations, especially when there are natural resources at stake. Historical resources like maps and treaties may be helpful, but still ambiguous or inconclusive, resources for resolving controversies. Consider disputes about the sky above a country, or the waters off its coasts. Deciding where exactly a country begins and ends may require a more exact definition of what a country is. Similarly, we would like to have an explicit definition of science in order to decide difficult cases. Everyone agrees that Galileo and Einstein were doing science, and that Luther and Gandhi were not. But if pressed to justify this sorting, how would we proceed? And what do we say about less clear-cut examples: was Newton doing science when he practiced alchemy? Were the Milesian cosmologists the first scientists, the first philosophers, or both? And there are ongoing controversies about the scientific status of "intelligent design" and string theory, as we shall see. History has sharpened our focus, but it's now time to seek a definitive conception of science.

Testability: the essence of science?

Too broad a definition would include activities that are clearly not science. For example, if we defined science as "the attempt to represent the natural world" then landscape painters and nature writers would be scientists. Perhaps we could narrow the definition to *mathematical* representations of natural systems since mathematics has been so crucial to the development of modern science. This might be too restrictive, however, since there are numerous putative sciences, such as archeology and geology, where mathematics plays a relatively minor role. You will find more figures and statistics in the sports pages of a newspaper than in a typical archeology journal. One important difference between a landscape painter and a geologist is that the latter aims to *explain* the structure of the landscape and not merely represent it as the painter does. And to do this the geologist will rely on various *theories* about the formation of the continents, the composition of soils and rock, climate variation, erosion, and so on. So, at a first approximation, we might define science as "the use of theories (sometimes mathematical in form) to explain natural processes and systems." And since we do not want to exclude sciences like psychology and economics, we should construe "natural" liberally enough to include systems and processes involving uniquely human behavior.

But this is still too broad, since the explanation of natural phenomena has not been the exclusive purview of science. As we noted in chapter one, prior to the emergence of science, the actions of gods and other primordial forces were routinely invoked to explain the origin and structure of the universe and catastrophic events like plagues and floods. Even today, it is not uncommon for the religiously inclined to explain their recovery from a serious illness by the theory that a powerful and benevolent god answered their prayers. But a medical scientist would explain the same event quite differently, perhaps by invoking immune system processes or the effectiveness of a drug treatment. So we might ask what distinguishes a scientific explanation of a given natural event from a non-scientific one?

The issue is not so much the truth or falsity of either sort of explanation. Maybe prayer, not the drug treatment, really was the reason for the person's recovery. Or vice versa. Still there seems to be a difference in the nature of the explanations offered, and this has to do with the means by which we could determine their truth or falsity. As we saw in the case of Halley's comet, one important kind of evidence for a theory is correct prediction. But in the present case it seems both theories correctly predict at least one relevant phenomenon. The recovery from the illness is to be expected on the theory of a powerful and benevolent god, and on the theory of an efficacious, properly administered drug. Since both theories are supported by this phenomenon, to decide between them we will need to consider which fares better when additional cases are taken into account.

Consider the drug theory first. Suppose we try the same treatment on several other patients with the same disease and have less success. This appears to count strongly against the effectiveness of the drug. Of course, a determined defender of the theory (perhaps a representative of the drug company) might suggest that there is some undetected difference in the additional patients that caused the treatment to fail – perhaps they had a different strain of the disease; or perhaps the dosages were off. But notice that these other factors can usually be checked directly. If it turns out that there simply are no relevant differences among the cases, then we can infer that the drug wasn't the cure in the first case. If it were effective in that case, there is no reason it wouldn't be in the others. So we would have strong evidence against the drug theory.

Applying the same methodology to the religious theory, suppose we found that other patients who prayed failed to recover. Does this mean the religious explanation is refuted? The advocate for this theory may first suggest, just as the proponent of the drug theory did, that there are undetected differences in the other cases that prevented divine intervention. As with the failed drug treatments, we can directly investigate this to some degree. We can check to see whether they all prayed for the same amount of time, in the same manner (though it may be hard to gauge sincerity and fervor), hold the same religious beliefs, and so on. But there is a crucial aspect of the explanation that we cannot check, namely the intentions and preferences of the divine being. The theory assumed that this being is benevolent and powerful, but it doesn't presume to know all of the being's motives. Perhaps there is some relevant difference in the patients who didn't recover that may not be evident to us, or that doesn't seem relevant to us, but is relevant to the object of prayer. With the medical theory the relevant mechanisms involved in the explanation are open to empirical check and control. But with the religious explanation there is no way to identify relevant factors in order to pinpoint the reason for the failed recoveries. We simply don't know what might matter to a god. Indeed, an imaginative advocate may attempt to turn the apparent evidence against the prayer theory to its credit by refining the theory. The reason only one patient recovered is because she prayed to a benevolent and powerful god who uniquely favors her. All of the evidence we have supports this refined version of the religious explanation!

The reader may begin to suspect that *nothing* can count as decisive evidence against the religious explanation: a persistent

TESTING THE POWER OF PRAYER

In 2006 a group of medical scientists, including Herbert Benson of Harvard University, conducted an investigation of the alleged medical power of "third-person" prayer (prayer on behalf of others). A large group of patients preparing to undergo heart surgery were randomly divided into three groups of approximately 600 each. For the members of one group, Christian volunteers prayed daily for a "quick, healthy recovery." Furthermore, members of this group were informed that they were being prayed for. The second group were also prayed for but informed only that this *might* occur. The third group were not prayed for but again told they might be. The study found no statistical difference in the speed or rate of recovery between the groups that were prayed for and the group that was not. Interestingly, between the two groups prayed for the study did find a slightly *worse* rate of recovery in the group that knew this. The authors of the study speculated that this might be due to a mild psychosomatic effect: "I must really be in bad shape if they're resorting to prayer!"

A religious critic of the study, Dr. Harold Koenig of Duke University, observed that it does not disprove the effectiveness of prayer because we have no way to predict how God will respond: "There is no god in either the Christian, Jewish or Moslem scriptures that can be constrained to the point that they can be predicted." In other words, a theory that makes no definite predictions is a theory that can't be refuted.

advocate will always be able to "explain away" failures or turn them into victories. According to the prominent twentieth-century philosopher Karl Popper, this is precisely why a religious explanation is not scientific. From a scientific point of view, the problem is not that nothing counts *for* such explanations but rather that nothing counts *against* them. They can't be *falsified*. So Popper offered the following "demarcation criterion" for distinguishing scientific from non-scientific theories and explanations:

> **Popper's Demarcation Criterion:** A theory is scientific to the extent that it is falsifiable by empirical tests.

Popper did not mean, of course, that in order for a theory to count as scientific it must be actually shown to be false – only that it must be *possible* to refute the theory by making "risky" empirical predictions or tests. Ideally, the theory will pass those tests (the predictions will be correct, like Halley's). If it does we can conclude that the theory is "corroborated," at least until the next attempt to refute it.

Popper developed his demarcation criterion while living in the intellectual hotbed of Vienna in the 1920s. He was very impressed with the recent empirical successes of Einstein's theory of relativity. Einstein's theory, both the special and general versions, had numerous surprising but testable consequences, for example that time would be measured differently on quickly moving bodies and that light would bend in the vicinity of very massive bodies. Since such predictions ran contrary to common sense and everyday experience, and since they could be derived with great mathematical precision, Einstein's theory qualified as highly falsifiable. Moreover, the predictions were corroborated by numerous and repeated experiments. And so Popper took the theory of relativity as his paradigm for genuine science.

But Popper was less impressed by other theories discussed in the coffee houses of Vienna, such as Freudian psychology and the Marxist theory of historical development. Apart from their potentially egregious socio-political implications, what bothered Popper most about these theories was that they both claimed scientific status. But in his mind they were mere "pseudo-sciences" because their proponents were unwilling or unable to specify the conditions under which they could be falsified. Consider first Marx's theory. Popper was dubious on grounds independent of falsifiability about the sort of historical laws posited by Marx: he thought historical change was subject to human free choice and therefore unpredictable. Still, he recognized that Marx's theory made predictions. It implied, for example, that the proletarian revolution required the conditions of advanced industrial economies. So the theory should predict the first revolutions in England or Germany. In fact the first major Marxist revolution occurred in Russia, whose economy was still primarily agrarian. But rather than dispense with their theory as falsified, many Marxists responded that the Russian Revolution was not really a communist revolution at all, but rather a "bourgeois revolution" like the French Revolution, or something else altogether. Similar apparent falsifications were dealt with in this ad hoc fashion. But this turned Marxism into an unfalsifiable dogma or, more charitably, a utopian dream: perhaps noble but not scientific.

Popper recognized that a respectable scientific theory could degenerate into dogma in the same way. Consider the popular nineteenth-century theory that a substance "phlogiston" is emitted in combustion. A prediction of this model should be that burning reduces weight. In fact it sometimes increases. To avoid falsification, some phlogiston theories proposed that phlogiston must have "negative weight." Such after the fact maneuvers to save the theory come at the cost of reducing its scientific integrity, according to Popper. Its defenders have

abandoned the "critical attitude" of true science. Popper emphasizes this in his major early work, *The Logic of Scientific Discovery*:
"those who uphold it dogmatically – believing perhaps that it is
their business to defend such a successful system against criticism
so long as it is not *conclusively disproved* – are adopting the very
reverse of the critical attitude that is in my view the proper one
for the scientist." As we shall see below, however, it is not
always clear when a failed prediction warrants abandonment of
a theory.

Some theories do not decline into dogma but are born
untestable. For example, Freud's theory had an even weaker
claim to scientific status than Marx's, according to Popper, since
it is immune to falsification from the outset. The reason is that
for any psychological or behavioral phenomenon Freudian
theory allegedly explained, it could also just as easily explain the
opposite phenomenon. Suppose we describe a fictional news
report of a man who attempts to drown a young child he doesn't
know. This man, a Freudian might speculate, suffers from
"repression" of his Oedipal complex. Now we reveal the truth:
this very same man actually risked his own life to save the child
from drowning. Ah, says the Freudian, the man had achieved
"sublimation" of his libidinous drives. It bears emphasizing that
Popper's point is not that Freudian psychology is false or useless.
He would not even deny that many people derive real benefits
from psychotherapy, just as from yoga or massage. Something
can be true and useful without meeting the condition of falsifiability required in science. It can also be false and deceptive. The
problem is, empirical testing can't reveal which it is.

Objections to falsificationism

Popper's demarcation criterion offers a clear and simple
condition for the scientific status of a theory: there must exist

MORE PROOF OF ASTROLOGY

Have you ever heard a weather forecast like this: "Partly to mostly cloudy with a chance of rain. Highs in the mid-70s to mid-80s?" From where I stand, the American Midwest in June, it's pretty hard to go wrong with a forecast like that. One of Popper's favorite targets was astrology. By making predictions about the course of human events, some modern versions of this ancient art give it a scientific veneer. But it also shields itself from falsification by making only predictions that are so vague that they're bound to be right. Allow me to illustrate. By a happy coincidence I happen to be writing this on my birthday (June 10). Since astrology is based in part on the positions of the planets in the sky at the time of birth, your horoscope ought to be most accurate if it is pegged to your birthday rather than merely the month-long range that determines your "sign." So I looked up an astrology website which provides a prediction about the year ahead "if today is your birthday." Here's mine:

"Relationships, mental pursuits, socializing, and creative expression are emphasized in your year ahead. Your ego is tied up in your social life and you have a chance to improve your social relationships, although a tendency to rebel could get you into hot water at times! A positive attitude and an enterprising spirit takes you places. Try to avoid making emotionally-driven decisions." (www.cafeastrology.com)

I don't have to wait a year to know this prediction is accurate! The only way it can go wrong is if I become a hermit and stop thinking or expressing myself creatively. The good news is I have a "chance" at social success (though I could also get into hot water). Note that even the bit of sage advice at the end is so banal that things can't really be too bad.

empirical tests that could decisively refute it. On this conception, scientists will be primarily occupied with efforts to falsify their favorite theories. This fits the popular image of science (and the self-image of many scientists) as highly critical and impartial. And certainly Popper's views have gained considerable currency among working scientists. On the whole, philosophers of science have been less convinced that falsificationism, at least in its simplest form, provides an adequate characterization of science. Popper's doctrine has faced a number of very serious objections (though I shall argue that in the end his core insight is right on the money).

To begin our critique of Popper – there is something at least odd in the notion that the aim of science is falsification. In Popper's view, it seems, we should be awarding the Nobel Prize to scientists who use severe tests to show us what the world is *not* like! The Popperian will reply that the aim of science is not falsification but the successful avoidance of attempted falsification, i.e. "corroboration." But what is so valuable about corroboration if we don't have reason to think that the most highly corroborated theories are true? In fact, on the Popperian view the best theories will tend to be *unlikely* since they make very specific predictions, just as a very specific weather forecast is more unlikely, all things considered, than a vague one. It is true that Popper tried hard to establish that science will tend to increase in verisimilitude or "truthlikeness." He attempted to show that verisimilitude will increase with falsifiability since both are functions of the content or information provided by the theory. However, such accounts of verisimilitude have faced insuperable technical difficulties. Like many others, Popper also drew an analogy between science and natural selection. But it is not clear why theories which prevail in the scientific "struggle for existence" will tend to be closer to the objective truth. Indeed, as we will see below, Popper had doubts whether the Darwinian principle of "survival of the fittest" was itself testable. (We'll return to the possible analogy between science and evolution in chapter four.)

Besides this philosophical defect, historians of science have been quick to point out that science simply does not work the way Popper envisions. Most importantly, otherwise valuable theories are not rejected simply because they have some false content. The Copernican model of the solar system championed by Galileo, though it had many advantages over the Ptolemaic model, was far from perfect. It seemed to imply that objects would be subject to an outward centrifugal force owing to the movement of the earth around the sun, like the force felt by children on a carousel. Galileo also retained perfectly circular orbits rather than Kepler's more accurate ellipses. These flaws implied some erroneous predictions but surely they did not warrant scrapping the entire Copernican system.

Or consider a contemporary example: the highly successful big bang model of cosmology. This theory correctly predicts the known rate of mutual recession of the galaxies, the existence and distribution of most of the natural elements in the universe, and the "cosmic background radiation" that echoes the big bang. It also meshes beautifully with the best theory of gravity and space, namely the general theory of relativity. But it also faces difficulties. For one thing, it seems to imply that the stars at the outer edge of galaxies should be much more dispersed from the center than they are observed to be. In other words, there does not seem to be enough matter within the galaxies to explain why they have "hung together" in the expanding universe. In the face of this anomaly, cosmologists have not abandoned the big bang model. Instead they have begun to speculate about "dark matter," the undetected gravitational glue that holds the galaxies together.

In neither of these cases, nor countless others, have scientists abandoned their theories in the face of anomalous observations. This would amount to "throwing out the baby with the bathwater." Since theories float upon a sea of anomalies, considered as an account of actual scientific practice it seems that falsificationism itself is falsified. Popper may reply, of course, that he is concerned with how science ideally *ought* to go, not how it sometimes actually *does* go. He may say that Galileo ought to have revised those aspects of the Copernican system that were problematic and that cosmologists should be actively pursuing alternatives to the big bang model rather than plugging its gaps with dark matter.

But there are two major difficulties with this sort of advice. First, alternative theories with comparable explanatory power are not easy to come by. There is no sense in abandoning a successful theory if you have nothing to replace it with, any more than in abandoning the only computer you can afford because it doesn't run every program you like. As Thomas Kuhn, an influential critic of the Popperian view of science, whose own concept of "paradigm shift" we will examine in chapter three, observes: "To reject one paradigm without simultaneously substituting another is to reject science itself. The act reflects not on the paradigm but on the man." A scientist who rejects a paradigm when first faced with empirical anomalies, Kuhn says, is like the proverbial carpenter who blames his tools.

A second problem with strict falsificationism is that sticking with an existing theory despite its flaws, even when this requires ad hoc repairs in the short run, can sometimes prove an excellent strategy over the long haul. For example, in the years following Newton's *Principia*, astronomers had difficulty fitting the orbit of Uranus to the path predicted by Newton's laws. Were the laws perhaps incorrect? Eventually it was speculated that the errant path was due to the gravitational influence of a hitherto unobserved planet or satellite. Using Newton's familiar laws, estimations were made of the likely sizes and orbits of such a body, and astronomers pointed their telescopes in the right directions. Eventually the culprit was discovered: the previously unknown planet Neptune. Popper pointed out that in this case the Newtonian theory was ultimately corroborated by the discovery of Neptune: it passed the test. However, the general point is that one cannot know in advance whether an inaccurate prediction is due to a problem with the theory, a problem with the measurement process itself, or simple ignorance of some other factor relevant to the phenomenon under examination. So judgments about the wisdom of sticking with an imperiled theory can only be rendered in retrospect. This is a point emphasized by another astute critic of Popper, Imre Lakatos: "Neither the logicians' proof of inconsistency, nor the experimental scientist's verdict of anomaly can defeat a research programme in one blow. One can be 'wise' only after the event."

The Neptune case points to another difficulty with the falsificationist ideal which is at bottom logical rather than merely historical. As Popper himself noted, when scientists test a theory by deriving a prediction from it, they rely on a host of assumptions or "auxiliary conditions." For example, in order to predict the orbit of Uranus, nineteenth-century astronomers needed to make various assumptions about the positions and masses of nearby bodies. They also had to rely on elaborate measurement devices and complex mathematical techniques and calculations. So when the predicted orbit was off, the scientists faced a choice of where to lay the blame: the assumptions about the other bodies, the calculations, or the theory itself. According to the philosophers Pierre Duhem and W. V. O. Quine, this choice is a matter of "convention" rather than logical principle. This is not to say, of course, that it is always reasonable to blame persistent predictive failure on inaccurate measurements. The church officials who attributed the apparent mountains on the moon to smudges in Galileo's telescope were being unreasonable even if they weren't making a logical error. However, the Duhem-Quine thesis does suggest that theory choice cannot be legislated solely in terms of consistency with known observations. Rather, it will depend on a complex "web" of considerations, some of which are emphasized by Popper himself, including novel predictive success, mathematical precision, simplicity, coherence with other theories, and promise of future application.

So there may be no single criterion for demarcating science from non-science or capturing the proper scientific attitude. The concept of "science" may in this way be similar to a concept like "game": there are many typical features of games – scorekeeping, rules, winners, and losers, etc. – but none of these are possessed by all and only games. This does not mean that it is a subjective matter whether something counts as a game or a science. We know, for example, that games must have intrinsic goals even if these vary greatly among and within games. Similarly, if a theory is absolutely immune to empirical challenge then it cannot be a serious contender in the scientific enterprise. And this I take to be the core insight of Popper. But although we will expect a science to involve empirically testable, mathematically precise, logically coherent explanations of natural systems, different sciences will exemplify these virtues in varying degrees.

MAYBE THERE'S SOMETHING TO ASTROLOGY AFTER ALL

In the 1960s the French statistician Michel Gauquelin published results of a large analysis of the birth-times of French children correlated with their later professions. Based on their precise location and dates, times of the birth are associated with positions of the sun, moon, and planets. While he found no particular connection between most of the planets and careers, in the case of Mars there appeared to be a correlation stronger than chance would predict. In particular, children who became eminent athletes were much more likely to be have been born with Mars near its culmination (highest point above the horizon). In a sample of 2088 sports heroes, 452 were born right after Mars rose or at its culmination. Chance predicts about 358. Statistically, the odds against this are roughly 5 million to one. This "Mars Effect" proved surprisingly robust when subjected to independent analysis. A long

MAYBE THERE'S SOMETHING TO ASTROLOGY AFTER ALL (*cont.*)

controversy ensued between Gauquelin and debunkers, with (not entirely unfounded) charges of duplicity from both sides. One important issue in the controversy has been the subjective element in the notion of "eminent." If Gauquelin "fine-tuned" his notion of eminence to fit the known data, perhaps by using different criteria for different kinds of athletes, he could ensure that an improbable number of "eminent" athletes turned out to have been born with Mars in the right place. The best way to prevent such gerrymandering – in any statistical study – is to stipulate the meaning of the variables in advance, and then look for correlations in future or in previously unexamined data.

Perhaps the most serious barrier to admitting the Mars Effect as a genuine scientific result is the lack of a mechanism that could explain it. Gravitation? But the doctor who attended my birth had a more profound gravitational influence on me than Mars. It's hard to see how such a weak force could determine my fundamental life choices. It's conceivable that people with genetics favoring athletic prowess also have genes that predispose them to be born with Mars at certain places in the sky just as the genes for blue eyes and blondness tend to associate. This eliminates the idea that planets seal our professional fate, but still requires a mysterious "triggering" of birth by Mars (and not by nearer bodies). Without a theory of the "Mars Effect" it is hard to know what would count for or against it.

Intelligent design

The difficulty of defining science has been underscored by many historical controversies about the status of various "fringe" sciences like alchemy, homeopathy, phrenology (psychology based on skull shape), and numerous forms of parapsychology.

Of particular interest recently is the debate over "intelligent design" (ID), which has been promoted in the United States as an alternative to Darwin's theory of evolution by natural selection. The ID movement gained prominence in the 1990s and early 2000s and was actually adopted into the science curriculum by a number of public school districts. In 2005, even the American President, George W. Bush, weighed in: "Both sides ought to be properly taught . . . so people can understand what the debate is about." In that same year, however, a federal court ruled that the teaching of intelligent design in public schools violated the U.S. Constitution's prohibition against "the establishment of religion." In arriving at his decision, Judge John Jones III (who had been appointed by Bush) relied on the testimony of philosophers of science about the definition of science. In his judicial ruling against ID he explicitly invoked a seemingly Popperian criterion of falsifiability: "once you attribute a cause to an untestable supernatural force, a proposition that cannot be disproven, there is no reason to continue seeking natural explanations as we have our answer."

Before considering whether Judge Jones was right in his Popperian repudiation of ID, it is worth noting a historical irony: Popper initially didn't think evolutionary theory itself qualified as science. His main worry was that the evolutionary principle of "survival of the fittest" was trivial and untestable. After all, what could it mean to be the "fittest" in the struggle for existence other than to survive (and produce the most offspring in the long run)? Then to say the fittest survive can only mean the survivors survive – and how could *that* be falsified? It therefore seemed to Popper that, just like Freud's theory, evolution could explain anything. If over thousands of generations a species develops a short tail, then that must contribute to its fitness – but likewise if it ends up with a long tail! Thus, Popper declared that evolution is not a testable scientific theory but rather a "metaphysical research program." However, he later recanted his dismissal of evolutionary theory when he recognized that biologists are able to specify independent criteria of fitness, over and above mere survival, in a given environment. To consider a simplified case, in very cold environments, all else being equal, species with warm coats and/or thick fat will be predicted to survive and reproduce better than those without. It will be difficult to actually test such a prediction, of course, since evolution works very gradually. However, biologists circumvent this problem by observing directly the evolution of species like drosophila (fruit flies) with very short life cycles and by relying on the long stretches of evolution recorded in the fossil record.

Turning to ID, it should be noted first that its proponents, such as the Lehigh University microbiologist Michael Behe, distance themselves from the biblical literalism of the earlier "creationist" movement. Indeed, Behe accepts the earth is very old and that evolution by natural selection does occur, at least at the "micro" or genetic level. But he maintains that the theory of an intelligent designer offers a better scientific explanation for the large differences between the species as well for the "irreducible complexity" of certain biological systems. What Behe means by an irreducibly complex system is one that cannot function successfully unless all its components are in place. He compares various microbiological structures, such as bacterial flagella, cilia, and the eyes to a mousetrap: if any of its parts are missing then it's useless for catching mice. The existence of irreducibly complex biological systems is allegedly a problem for classical evolutionary theory. Since there is no advantage to having only parts of these systems they would not evolve gradually: it's very useful to have eyes but having only a lens or a retina or a cornea does me no good. The argument, which is really just a biochemical version of natural theology's "argument from design," has been addressed in great detail by defenders of evolution (see Further reading).

They have maintained that the systems identified by Behe are not irreducibly complex, that Behe relies on a simplistic conception of the "parts" of biological systems, that features originally selected for one advantage can come to serve a different function, and so on.

Critics of ID charge that it relies on the fallacy known as "false dichotomy:" even if classical evolution cannot explain irreducible complexity it doesn't follow that these systems were designed – there may be other explanations besides just these two. In fairness, one can't require from ID proof that it is the only possible explanation for complexity: no scientific theory could meet such a demand. Yet in order to be taken seriously ID must at least present an alternative explanation for complexity, and a better one than evolution. Unfortunately, ID proponents have had more to say about the supposed weaknesses in evolutionary theory than the details of their own theory. Near the conclusion of his book *Darwin's Black Box*, Behe asserts in his own defense: "The conclusion that something is designed can be made quite independently of knowledge of the designer." Of course he's right that someone can draw whatever sorts of conclusions they want from the evidence. The question is whether the hypothesis of an intelligent designer is *justified* by the evidence. And we will see that this question is very hard to answer without independent knowledge of the designer.

How do we test the theory that complex biological systems were designed by an intelligent agent? We would normally test a theory by deriving a prediction from it about past or future phenomena, or about an experimental outcome, and then check the result. But nothing specific follows from the hypothesis of an intelligent designer about what we should observe in nature or an experiment, since we simply don't know the agent's intentions or skills. Perhaps the agent would make eyes and flagella, or perhaps they would make entirely different things that appealed to their unique tastes. It is impossible to say whether flagella count for or against Behe's intelligent agent since, for all he has told us about this agent, it might well despise such systems. Without more detail, all that follows from the supposition that the agent exists is that biological systems *might* exist (if the agent desires to make them) and they *might* be complex (if the agent prefers complexity). There is no real test of such a supposition because whatever we could observe is perfectly consistent with it.

But maybe we can say more about what an intelligent designer's intentions and skills might be, and thereby derive more specific predictions from the theory. Perhaps we should expect that an intelligent designer would produce biological structures that serve the interests of the organisms that possess them (just as natural selection predicts the gradual evolution of adaptive traits). The eyes of certain fishes, for example, have just the right structure to facilitate visual perception in the cloudy waters they inhabit. The problem is that this very design is contrary to the interests of smaller species of fish who are spotted and preyed upon. As Tennyson observed, nature is so "red in tooth and claw" that it is difficult to conceive it as the product of a benevolent designer. Furthermore, there seems to be an abundance of manifestly poor design in nature, assuming the designer's intention was to further the interests of the designed. For most species of insects, fishes, and birds, early death is the norm and survival (let alone reproduction) is the rare exception. Even for most humans, the sad fact is that life is a daily struggle for food and shelter, with the persistent threat of disease and violence. In the face of this apparently unintelligent or malicious design, the proponent of ID may want to invoke the familiar caveat that the designer's plan is beyond our understanding. But this is merely to revert to the untestable proposition that everything complex is designed.

THE FLYING SPAGHETTI MONSTER

Although Michael Behe suggests that modern molecular biology is the lynchpin of ID, the inference from biological complexity to an intelligent designer is a staple of the very old tradition of "natural theology." In the classic work in this tradition, *Natural Theology; or, Evidences of the Existence and Attributes of the Deity*, William Paley argued that the human eye implicates a designer no less than does a ticking watch found on the seashore. The objection raised above – that the designer hypothesis lacks sufficient detail to test – is just as old. In his *Dialogues Concerning Natural Religion*, David Hume pointed out that design evidence is neutral between the traditional Western hypothesis of God and the "Brahmin" view that the world arose "from an infinite spider, who spun this whole complicated mass from his bowels, and annihilates afterwards the whole or any part of it, by absorbing it again, and resolving it into his own essence." The same ironic strategy of offering exotic alternatives to the favored design hypothesis has been adopted by contemporary opponents of ID, who have proposed that biological complexity is evidence for a "flying spaghetti monster" and have lobbied for instruction in this theory in American schools.

String theory

Philosophical debates about the nature of science arise frequently and naturally within science itself, especially when longstanding theories are under assault and radical alternatives are proposed. As we noted in chapter one, the conservative critics of Galileo argued that it is not the business of science to speculate about the "true causes" of astronomical phenomena – it is enough to "save the appearances." Einstein raised exactly the opposite complaint against the early quantum theory, which despite its empirical successes he found to be "incomplete" due to the statistical nature of its laws. (This is the source of his famous quip that "God does not play dice with the universe.") As Thomas Kuhn has observed, during these times of "crisis" scientists often turn to philosophical analysis: "It is no accident that the emergence of Newtonian physics in the seventeenth century and of relativity and quantum mechanics in the twentieth century should have been preceded and accompanied by fundamental philosophical analyses." Very recently the issue has been joined once again at the frontiers of modern particle physics, and the controversy has focused squarely on the question of testability.

The two greatest theories of twentieth-century physics, quantum theory and relativity theory, each demanded a dramatic revision in Newtonian or "classical" physics. In quantum mechanics, parameters like momentum and position are not fully determinate independent of measurement, and separated but "entangled" systems seem to exhibit a mysterious non-gravitational action at a distance. In relativity theory, time and space are not absolute quantities, as Newton supposed, but instead variables dependent on motion and mass. Despite these surprising features, both theories have been enormously successful in their respective domains. Applied to the "small-scale" behavior of atomic and sub-atomic particles, quantum theory has accounted for various kinds of atomic, electromagnetic and even chemical phenomena. Relativity, especially the "general" theory which deals with space and gravity, has been applied with equal success to "large-scale" sciences like cosmology and the astrophysics of stars and black holes.

The problem is that there is no obvious way to "unify" these two theories in contexts in which they ought both to apply. For example, according to the so-called "standard" model of elementary particles (those from which atoms and molecules, and so everything else, are composed) there are a number of extremely tiny particles (quarks, leptons, electrons, and so on) governed by four fundamental forces: the strong and weak nuclear forces, electromagnetism, and gravity. The trick is accounting for these forces in quantum mechanical terms. There has been some progress in explaining how the first three forces might arise from quantum processes. But it has proven much more difficult to find an adequate theory of "quantum gravity."

Fundamental to quantum theory is the assumption that waves can be described as particles and vice versa. So there must be "gravitons" associated with the gravitational waves predicted by relativity. One might attempt to explain the behavior and interactions among gravitons, and with electrons and other particles, in terms of their characteristic wavelengths and frequencies. Instead, string theory reduces all such processes to the "vibrations" of one-dimensional strings. The strings curve and vibrate according to the energies involved, and join and split as elementary particles interact and merge. For example, the weakest force, namely gravity, has the least amplitude in the vibration of its string. Using various mathematical techniques, such as "renormalization," string theorists have developed a consistent way of representing quantum gravity within the standard model and a unified description of the elementary particles and forces in terms of strings.

String theory thus offers a mathematically powerful and coherent method for unifying quantum and relativity theory, a problem that dogged theoretical physics for much of the last century. But it faces many difficulties. For one, the theory seems to require the existence of many more than the three dimensions of space that we observe, perhaps as many as eleven or even twenty-six dimensions. But in itself this curiosity should probably be taken with a grain of salt. After all, quantum theory and relativity already require drastic revision to our common sense conceptions of time and space. Is it any surprise that their unifi-cation would have additional counter-intuitive consequences? As the history of physics has shown us, common sense frequently gives way to scientific innovation.

More seriously, string theory can be formulated using numerous different dimensionalities and there is no obvious way of choosing between them. Even if we restrict ourselves to the "simplest" spaces, which seem to be nine-dimensional (plus one time dimension), there are hundreds of logically distinct but internally consistent ways of representing the fundamental forces and particles as strings. These are not mere differences in mathe-matical formalism: the different theories posit very different underlying physical realities. This is a familiar enough scenario in science. We have seen for example that the Ptolemaic and Copernican models (not to mention the "hybrid" of these developed by Tycho Brahe) can be shown to account for the same appearances in different ways. The special difficulty here, unlike in the astronomical case, is that it is hard to envision a "crucial experiment," or indeed any experiment at all, that would favor one model of string theory over another. Indeed, however mathematically elegant and unifying, string theory seems to have no consequences that can be experimentally tested. The main problem is that the extra dimensions posited would need to be extremely short, at the so-called "Planck length" of quantum theory, in order to imply the "quantization" of the gravitational force. But, as Harvard string theorist Lisa Randall recently conceded, the detection of such lengths would require energy levels "ten thousand trillion times the reach of current particle accelerators."

This state of affairs has generated considerable controversy in the world of theoretical physics, much of it focused on the question of string theory's testability. For example, the Nobel Prize winning particle physicist Sheldon Glashow recently commented:

The string theorists have a theory that appears to be consistent and is very beautiful, very complex, and I don't understand it. It gives a quantum theory of gravity that appears to be consistent but doesn't make any other predictions. That is to say, there ain't no experiment that could be done nor is there any observation that could be made that would say, "You guys are wrong." The theory is safe, permanently safe. I ask you, is that a theory of physics or a philosophy? (Nova Interview, PBS)

Supporters of string theory responded that despite its weaknesses, their account of quantum gravity is the only, or at least the best, game in town. However, the question is not whether string theory is the best scientific explanation available, but whether it qualifies as a *scientific* explanation at all. As the prominent string theorist Edward Witten has observed, string theory certainly predicts and explains the force of gravity – that's what it was invented for after all. Yet, the mere fact that a theory predicts a certain phenomena does not ensure the theory is testable. As we have seen, ID predicts the existence of complex phenomena and my horoscope predicts that next year I will "have a chance to improve my social relationships." But these predictions cannot go wrong, which is precisely Glashow's point about string theory. All that the many versions of string theory currently predict is just what is already known according to the standard model of particle physics. In the same way, ID predicts the complexity of the eye, the flagellum, etc; but so do the theories of the "Brahmin's spider," the flying spaghetti monster, and so on.

In the end, can fields like string theory at the cutting edge of mathematical physics be distinguished from alternatives to mainstream science like ID? The answer, I think, is clearly yes. First of all, although neither string theory nor ID can be currently tested, the proponents of string theory have indicated various ways in which it is *testable* and falsifiable. They note, for example, that string theory requires the existence of "supersymmetric" twins of the known particles. Because they are probably very heavy, detection of these "superparticles" would require energy levels far greater than what is attainable in existing accelerators. But they may soon be found using the new Large Hadron Collider (LHC) constructed under the France-Switzerland border. This would not be the first time scientific confirmation (or falsification) had to wait on theoretical and technological advances: Copernicus' heliocentric model and Einstein's general theory of relativity are other examples. Competing models of the long-term fate of the universe – perpetual expansion vs. eventual contraction – are not easily testable at present nor are certain interpretations of quantum mechanics such as the so-called "many-worlds" interpretation.

Unlike ID, however, these exotic theories characterize physical scenarios in sufficient detail, and in terms sufficiently related to familiar physical concepts, that it is possible to specify what kinds of observations and data would bear on them. For example, it had long been clear that detection of the gravitational red-shift among galaxies would provide information about the rate of the expansion of the universe, and that this was directly relevant to competing models of the ultimate fate of the universe. These observations finally became possible in recent years, using the Hubble Space Telescope, among other means. But it is impossible to imagine what observations, using any powerful technology, would confirm or disconfirm an intelligent designer. The claim that there is such a being in itself says nothing about any physical process, however difficult to detect.

Second, despite the acrimonious volleys from opposing sides of the string theory debate, all agree that the question must eventually be settled by standard experimental means, presumably using data from the next generation of high-energy colliders. Contrary to Glashow's remark above, string theorists such as Michael Hughes of Rutgers University have suggested numerous technologically feasible experiments that could disconfirm or falsify string theory. By contrast, the proponents of ID have been less forthcoming about what sort of future data or experiments would disprove or mandate revision of their theory. In a word, ID lacks the "critical attitude" Popper finds at the heart of science.

Third, string theory, even if it cannot ultimately be tested in a decisive fashion, is grounded on theories that are clearly testable and successful. But the ID hypothesis, to the extent it has been elaborated at all, seems to be derived from a particular religious tradition and grounded on anthropomorphic intuitions about the values and intentions of intelligent and powerful beings. Such traditions and intuitions are worthy subjects of historical and philosophical investigation – they may even be valid in some sense – but they aren't matters for empirical testing.

There are numerous philosophical problems involved in marking out the boundary between science and other forms of human inquiry. Some of these will be taken up in the following chapters. But we have found in this chapter that empirical testability is a fundamental prerequisite for science. This is not to say that empirical tests are all that matter or that they must be decisive in all circumstances. It will often make eminent sense to stick with a theory that has failed an important test, or for which no feasible tests currently exist. But a theory that says nothing definite about the observable world at all, or (what amounts to the same thing) says about the observable world only that it is however we find it to be, is not scientific. The demand for empirical testability reflects the ideal that the ultimate arbiter in science is not faith or utility or logic, or even truth, but the empirical world itself.

3

The scientific method

The previous chapter revealed that science is distinguished more by its method than its subject matter. The natural world is the subject of landscape painters and poets no less than of botanists and astronomers; but the latter investigate nature by means of empirically testable theories and explanations while painters and poets rely on perspective, impression and metaphor. This difference in method is perhaps sufficient to capture the essential demarcation between science and non-science. But it is time to form a more definite conception of the scientific method. Given scientific knowledge is based on testable theories, when exactly can a theory be said to have passed (or failed) a test? And at what point can we infer from passed (or failed) tests that a theory is definitely true (or false)? Can we have reason to conclude a theory is at least probable, based on its tests, even if we can't be sure it's true? And how do we arrive at the theories we test in the first place? In order to come to terms with these issues, it is helpful to introduce a broad distinction between two historically prominent approaches to the problems of method: deductivism and inductivism.

Deductivism vs. inductivism

In the most general sense, reasoning (whether scientific or otherwise) involves drawing conclusions about a subject from some given information. To draw conclusions without relying on relevant information is mere guessing. Reasoning is good when the information relied on strongly supports the conclusion, and bad when the information supports the conclusion only weakly or not at all. Notice that the quality of reasoning is a matter of the *connection between* the information and the conclusion, not the independent truth or falsity of either. So it's possible to reason well to a false conclusion (or poorly to a true one). If someone predicted a continuing rise in gas prices, based on increased consumption, just prior to an unexpected discovery of massive oil fields, they made a reasonable conclusion and were simply unlucky.

Logicians draw a fundamental distinction between two forms of potentially good reasoning: *deductive* and *inductive*. With good deductive reasoning, if the information relied upon is true then that *guarantees* the truth of the conclusion. Here's an example:

1. All penguins are birds.
2. All birds are animals.
3. So, all penguins are animals.

Since the first two statements are true the conclusion must be true as well. Such arguments are said to be deductively *valid*.

Good inductive reasoning does not guarantee the truth of the conclusion based on the premises, only makes the truth *likely*. Here's an example:

1. Hundreds of species of penguins have been observed so far and all are swimmers.
2. So, all species of penguins are swimmers.

Assuming the first statement is true then the conclusion is probable, but not absolutely certain. It is possible that some remote and undiscovered species of penguin can't swim. Inductive reasoning, unlike deductive reasoning, is said to be *ampliative*, meaning the conclusion "goes beyond" the premises. In the argument above, a conclusion is drawn about all penguins (including those that may be observed in the future) from the

ones so far observed. It is because inductive reasoning is ampliative that it is riskier than deductive. An inductive argument in
which the risk is small, because the facts appealed to really do
make the conclusion likely, is said to be *strong*.

Deductive reasoning is characteristic of fields like mathematics and pure logic: the conclusion of a proof in Euclidean
geometry that the sides opposite equal angles of a triangle are of
equal length, is supposed to follow with certainty from Euclid's
axioms, not with mere probability. Inductive reasoning is
characteristic of fields like polling, where from a representative
sample of a population we infer something about the population
as a whole with some degree of confidence but not with
absolute certainty. The question that will concern us shortly is
whether scientific reasoning generally, i.e. the scientific method,
is primarily deductive or inductive in form.

But before turning to this question, it is worth noting that
associated with the opposing views on this question are opposing views about the *origin* or basis of knowledge. Because they
emphasize purely logical relationships among information,
deductivists about scientific reasoning have historically tended to
support *rationalism*. This is the view that most or all knowledge
is based on pure reason or intellect. Most inductivists, because
they stress fallible reasoning from available experience, tend to
support some version of *empiricism*. This is the view that most or
all knowledge is based on experience or observation rather than
pure reason. It should be said, however, that pure rationalism
about science is rare today and many deductivists about scientific
method embrace a form of empiricism about the source of
knowledge, especially recently: Popper and his followers are
examples. But nearly all inductivists are empiricists.

Many of the ongoing debates about scientific method have
their origin in the Scientific Revolution, when philosophers
attempted to formulate clear rules to guide the development of
the new science. The opposing deductivist and inductivist

conceptions of method are exemplified by two giants of this era: Descartes and Newton. Descartes' earliest work was in geometry – hence the "Cartesian coordinates" of analytical geometry – and he carried the deductive orientation of mathematics into his mature work in philosophy and science. Dissatisfied with the uncertain views he had received from education and sense experience, Descartes employed skeptical arguments to demolish these former opinions. How can I be sure that what I now perceive is not a dream, or a deception conjured by a malicious demon? The aim of such fantastic hypotheses – the method to this madness – was to uncover absolutely certain foundations upon which to rebuild science.

Eventually, Descartes discovered one proposition that must be true even when dreaming or under a demon's control: "I exist" (the famous "I think therefore I am"). From this proposition, along with certain allegedly self-evident facts about the concepts of God and causality, Descartes attempted to deduce all the truths of physics, astronomy, optics, and even biology and medicine. Thus he boasts in the preface to his major scientific work, the *Principles of Philosophy*: "there has been no one up until now who has recognized the principles which enable us to deduce the knowledge of all the other things to be found in the world." Near the end of the same text, he declares that his demonstrations are "so certain that even if our experiences seemed to show us the opposite, we should still be obliged to have more faith in our reason than in our senses." So for Descartes deductive reasoning is the preferred method of science because it provides more certainty than can be obtained by inductive reasoning from experience.

Although Isaac Newton was an even greater mathematician than Descartes, he regarded experience as playing a more important role in science than pure reason. Indeed, he was very suspicious of theoretical notions or "hypotheses" that went beyond observable phenomena. The reason he pledged in the *Principia* to "feign no hypotheses" was because "hypotheses whether physical, or based on occult faculties, or mechanical, have no place in the experimental philosophy. In this experimental philosophy, propositions are deduced from the phenomena and made general by induction." The only role for "deduction" is simply in describing a particular phenomenon. After that the description is "made general by induction." For example, I might deduce the particular rates of fall of various bodies near the earth by carefully measuring them. But then I inductively generalize a law for all falling bodies (Galileo's law) and finally for all mutually attractive bodies whatever (law of universal gravitation). Newton recognized that generalizing from finite observations to a universal law involved uncertainty and risk; but he thought this was the best we could do if we hoped to have a science grounded firmly on empirical evidence rather than metaphysical hypotheses: "In experimental philosophy, propositions gathered from phenomena by induction should be considered either exactly or very nearly true."

In their own scientific work, neither Descartes nor Newton adhered strictly to their official methodologies. Descartes conducted many experiments in his far-flung investigations in mechanics, optics, and physiology. His considered position seems to have been that although science should begin from metaphysical first principles, it needed to be tempered by observation, especially when dealing with very detailed and specific phenomena: "the further we advance in our knowledge the more necessary they [observations] become." Likewise, Newton's abjuration of hypotheses that go beyond experience needs to be considered in relation to the hypothetical nature of much of his own science. The universal force "gravity" was itself a hypothetical entity, whose mysterious ability to act at a distance bothered Newton to such an extent that he tentatively introduced another hypothetical entity, the all-pervasive material "ether," as the medium of gravitational attraction.

Furthermore, the most powerful tool in Newtonian method was a deductive one, the calculus for representing complex and continuously varying systems.

But although it is likely that in practice science will involve a mix of deductive and inductive reasoning, there remains a real and important difference in the two approaches. Newton insisted that the hypothesis of gravity, along with the laws of nature, are ultimately inferred inductively from particular observations rather than deduced a priori in the geometrical style of Descartes. He considered this the key to keeping theoretical science "honest" or firmly rooted in experiment. Unfortunately, Newton did not say a great deal about exactly how scientific theories are induced or "gathered" from the phenomena. So it is worth considering briefly two more detailed philosophical theories of induction which exerted enormous influence on science right through to the twentieth century.

The first great inductivist philosopher of science was Francis Bacon. An Elizabethan nobleman, and contemporary of Shakespeare, Bacon devoted himself to systematizing the new sciences that were emerging in the sixteenth century. For Bacon, science offered not only the intellectual delight noted by Aristotle, and evidence of divine wisdom prized by the natural theologians, but also the potential to harness the power of natural forces: "human knowledge and human power meet in one; for where the cause is not known, the effect cannot be produced." But to attain this knowledge and power, science needed to be pursued in a systematic and progressive manner. This meant first of all evicting from our minds certain "idols" that divert and mislead our thoughts. For example, "idols of the tribe" (prejudices endemic in human nature) cause us to measure all things by human standards, while "idols of the theatre" (slavish deference to authority) make us place undue trust in eminent teachers and philosophies (e.g. Aristotle). For Bacon, the only true authority in science is direct and unprejudiced observation.

In his work *Novum Organon* ("New Method") Bacon calls for less reliance on the traditional Aristotelian syllogisms (deductive logical forms) and more on induction, of which he complains, "the logicians seem hardly to have taken any serious thought." Furthermore, he proposes to replace what he calls the "method of anticipation" which "flies from the senses and particulars to the most general axioms" with his own more cautious and deliberate "method of interpretation" which "derives axioms from the senses and particulars, rising by a gradual ascent so that it arrives at the most general axioms last of all." The elaborate inductive process he prescribes begins with the recording of unmediated experiences, which are then organized and tabulated by qualitative and quantitative criteria. From these tables, more general "axioms" are abstracted, perhaps with the aid of additional experiments when inconsistent observations are encountered. From these mid-level axioms, the researcher eventually ascends to the most general level, the laws of nature. So one might begin by measuring and cataloging the rate of fall of various objects from different heights and in various media. This may lead to the "axiom" that rate of fall is reduced proportionally to the surface area of the falling object and the density of the medium of fall. These axioms may in turn suggest experiments involving different weights but similar surface areas in the same media, and lead finally to the discovery that all bodies fall in a vacuum proportionally to the square of the time they fall (Galileo's law).

Although Bacon predicted that science could be brought to completion within "a few years" by conscientious employment of his inductive machine, there are good reasons to question such enthusiasm. First, it seems unlikely that the unguided, ground-level collection of facts can lead anywhere interesting. Consider again the example of falling bodies. The way I described the process, not *all* facts were cataloged – there was no consideration of the color or age of the falling bodies nor the race or gender of the people who dropped them. The reason we ignore such facts is presumably because we gather our information with a particular problem in mind (the rate of free fall) and a host of tacit assumptions about what is and is not relevant to its solution (the shape of the object is likely to be relevant but not the color). It is hard to see how absolutely undirected fact-gathering would lead to anything but a hodge-podge of unrelated trivia.

Second, even if we could compile and organize all the infinite observable facts, and derive from them the most general laws or regularities, this would not give us all we want from science. One of the laws obtained in the medical domain might be that people with shingles had invariably had chicken pox in the past, though chicken pox only rarely led to shingles. A shingles sufferer might reasonably ask medical science for an explanation of why she contracted shingles while her equally spotted siblings escaped. There is a complicated answer to this question, involving genetics and details about the immune and nervous systems. But they are not available to the strict Baconian since he is concerned solely with the organization of facts and not with their unobservable cause or explanation.

Finally, even if theoretical explanations are somehow intro-duced into Bacon's method, as Bacon himself sometimes suggests, these explanations do not seem to be straightforwardly derivable by "interpretation" from the collected facts. The partial association of two familiar diseases allows countless explan-ations, just as a finite number of data points on a graph can be "fitted" to an indefinite number of different curves. Of course, experiments may help to distinguish among the rival interpreta-tions; but only once the rivals have been proposed. The point is that the facts themselves, however judiciously organized, do not generate their own explanation. (Incidentally, Bacon was an enthusiastic experimenter himself and is said to have expired from a cold he caught stuffing a chicken with snow as an experiment in refrigeration.)

Inductive theories of scientific method were refined significantly as modern science progressed, especially by nineteenth-century figures like William Whewell, William Herschel and John Stuart Mill. Here I will briefly examine Mill's rules of induction, since their power and clarity have made them a touchstone for all subsequent inductive systems. Mill was an English polymath, perhaps best known for his utilitarian moral philosophy and libertarian political theory. But his systematic account of induction, presented in the two-volume *System of Logic*, has also been extremely influential in the philosophy of science. Mill aimed to provide the same footing for inductive inference as for deductive sciences like geometry: "to reason is simply to infer any assertion from assertions already admitted and in this sense induction is as much entitled to be called reasoning as the demonstrations of geometry." Along with analysis of traditional syllogistic logic, the nature of language, and the classic fallacies, Mill identified four "methods of experimental inquiry":

- The method of agreement
- The method of difference
- The method of concomitant variations
- The method of residues

We will discuss the first two methods to get a sense of the main strengths (and weaknesses) of Mill's inductive system since the second two are in essence merely extensions of these.

First, the method of agreement. Suppose I am concerned about the irregular and extremely irritating behavior of my cat. An otherwise pleasant creature, she occasionally spends a good part of the morning prancing about the house yowling. As a good empiricist I begin to keep a log of these mornings, noting the weather, household activities, the cat's breakfast, the behavior of my other cat, and so on. Failing to discover any association between these factors and Kitty's erratic behavior, I eventually realize she always acts up on Wednesday mornings – garbage day.

Eureka! I infer that her irritation is caused by the garbage and recycling trucks that parade noisily through my neighborhood every Wednesday. By observing numerous instances of an effect I have been able to identify the one thing they all have in common, or agree on, and so I infer that this thing is the cause. As Mill puts it: "If two or more instances of the phenomenon under investigation have only one circumstance in common, the circumstance in which alone all the instances agree, is the cause (or effect) of the given phenomenon."

Now consider the method of difference. Suppose I would like to know why one of the eight tomato plants I planted this spring is dying while the others are thriving. The obvious thing to look for is something special about this plant. I know they are all the same variety of tomato and were purchased at the same garden store. I can't recall treating (or mistreating) this plant differently from the others – it's been given the same amount of plant food and water. I notice, however, that the struggling plant is the only one that borders a public pathway. Since my neighbor walks his beagle along this path every evening I begin to suspect that this plant has been subject to excessive "fertilization" courtesy of the beagle. When this is confirmed by diligent observation, I infer that the beagle's marking is indeed the cause of my tomato plant's demise. Thus, by observing a number of cases I can detect the cause of some special feature of one case by isolating some other unique difference it has from the others. As Mill puts it: "If an instance in which the phenomenon under investigation occurs, and an instance in which it does not occur, have every circumstance in common save one, that one occurring only in the former; the circumstance in which alone the two instances differ, is the effect, or the cause, or an indispensable part of the cause, of the phenomenon." Thus stated, the method of difference leaves open which of the circumstances is the cause and which is the effect. Temporal priority and background knowledge can help distinguish cause from effect.

In the present case, we can safely assume that decline of the plant followed, and so did not induce, the beagle's marking behavior.

Mill's methods capture a powerful and natural way of reasoning about causes from observation, especially when we combine them in what Mill calls the "joint method of agreement and difference:" by looking for agreement we rule out spurious factors and by looking for difference we narrow in on the decisive factor – the "smoking gun." If we can exclude all the factors that we know are not causally relevant because the effect actually occurs without them (no agreement) and then systematically eliminate the remaining factors that aren't necessary or sufficient for the effect (no difference) we will eventually arrive at the actual "difference-maker." Suppose in the case of my irritable cat I initially use the method of agreement to find that there are several factors that have accompanied her bad mornings: garbage day, the absence of my daughter, and dry cat food for breakfast. Then, by the method of difference I can rule out the breakfast: since she has that every weekday this does not mark out a difference from the days in which she isn't irritable. On the other hand, my daughter Sophie is only gone on Wednesdays. Fortunately, I can pursue the method of difference further by arranging for her to stay home some Wednesday. If the cat is still irritable on that day then I know the garbage truck is the difference-maker (assuming I haven't missed a relevant agreeing factor). If the cat is not irritable then I know Sophie's absence is the cause (assuming I haven't missed some relevant other factor that disagrees in this case).

This example confirms two important observations made by Mill:

(i) "Of these methods that of difference is more particularly a method of artificial experiment." For what has so far agreed with the effect is simply what we learn from past experience. But using the method of difference we intervene in future circumstances by varying the plausible causal factors, in hope of singling out the true difference-maker.

(ii) It is "by the method of difference alone that we can ever, in the way of direct experiences, arrive with certainty at causes." By direct observation the method of agreement may only manage to reduce to several the number of circumstances that always accompany the effect, but the method of difference applied rigorously will allow us to identify the unique circumstances needed to produce the effect. For these reasons the ideal procedure is to apply both methods.

But the conditions necessary to apply Mill's methods jointly can be exceedingly hard to achieve in actual scientific investigation. The phenomenon under investigation may be relatively rare or beyond our ability to affect. For example, it is hard to see how one could single out by observation the unique difference that accompanied all of the appearances of Halley's comet. And likewise if we are dealing with highly complex biological systems, like the human body, it will be very difficult to isolate the difference-maker with respect to a certain effect. For example, a huge array of inter-dependent muscular, chemical and electrical processes in far flung parts of my body "agree" with each of my heartbeats; but how does one determine which of these (or which combination of these) are the essential factors? And it is always hard to know whether we have really identified by observation and experiment all the possible relevant agreeing and differing factors to begin with, as the cat example illustrates. Or consider a problem from economics: what caused the stock market crash of 1929? Since it only happened once, there is no agreement to find, and since the past is foreclosed to our intervention we can't vary the circumstances systematically to determine under what circumstances it would not have occurred. We can look for what was "different" in the events immediately

before the crash, but there will be many of these that are potentially relevant, such as government policies, banking activities,
national and international politics, and so on.

Mill's methods are best approximated in the "controlled"
experiments often employed in biological and medical sciences.
In such studies, a sample population is randomly separated into
two groups so that they are as similar as possible. The suspected
causal factor is then introduced into one of the groups, the
"experimental" group, while a placebo is provided to the other
group, the "control" group. If the incidence of the effect is
significantly greater in the experimental group, we know the
suspected cause is the real difference-maker since it is "the
circumstance alone in which the two instances differ."
Controlled experiments are sometimes financially or morally
prohibitive (e.g. if the effect in question is a medically bad
one) so researchers must resort to less powerful strategies such as
versions of the agreement method (e.g. surveying those who
already have the effect, hoping to find something they have in
common).

There is a more fundamental problem with taking Mill's
inductive methods as the standard for scientific inquiry, which
also plagues Bacon's inductive machine. In order to effectively
utilize the methods for finding difference-makers we must be
guided by some theoretical or background assumptions about
what is likely to be the causal factor. Even in the very simple
example of my irritable cat, there are countless experiments I do
not undertake because I am confident of their irrelevance to her
Wednesday morning behavior: I don't shift my Tuesday poker
night, cancel my "word-of-the-week" e-mail subscription, ask
my neighbor to skip his mid-week dinner out, and so on. In
order to avoid an enormous search through all the logically
possible circumstances that agree with the effect, I restrict myself
to those which are plausible given what I already know about
cats, irritation, causality, and so on. Such assumptions seem

necessary even when conducting controlled experiments. If we are searching for an effective treatment for a chronic disease, we don't try every conceivable drug and therapy, but are guided by previous results, accepted theory and common sense.

Theory is especially vital to guide experimentation in highly advanced sciences like particle physics or astronomy. In an effort to identify the cause of supernovae it would be futile (and expensive) to scan the sky with our most powerful telescopes searching for agreements and differences in the stars. We might notice patterns and correlations of various kinds, but this would not get us very far in determining the cause of supernovae. Part of the problem in this case is that we can't fully employ the method of difference because we can't manipulate the stars. We can use accelerators to manipulate sub-atomic particles, yet it would be equally futile to run experiment after experiment hoping to find the difference-maker that reveals the precise structure of matter. In both of these cases, the experiments we attempt, the technologies we use, and our interpretation of the data are all constrained by detailed background theories. Mill's methods may be useful tools in daily life, as a way of guiding the early stages of a causal investigation, but they can have only a minor role in the most advanced sciences and must be supported by pre-existing theoretical conjectures.

Mill acknowledged that hypotheses could offer "temporary assistance" to induction from experience when the domain of inquiry is highly unfamiliar. But he insisted that the hypotheses themselves must ultimately be reduced to a form that can be directly verified by the method of difference. Mill's opposition to hypothetical reasoning, reminiscent of Newton, was severely criticized by his countryman William Whewell as inverting the dynamics of actual science by making hypotheses secondary and eliminable. According to Whewell, who was a dabbler in numerous sciences and an avid historian of science, hypotheses (conceptions) are needed at the beginning of inquiry in order to give meaning and structure to the facts: "To hit upon a right conception is a difficult step; and when the step is once made the facts assume a different aspect from what they had before. . . Before this the thoughts are seen as detached separate, lawless; afterwards they are seen as connected, singular, regular." Whewell's point is that unless the scientist is permitted to introduce radically new ideas, as has so often occurred in the history of science, the old facts will remain jumbled and puzzling.

Mill objected that the conjectural approach of Whewell, in allowing hypotheses that extend beyond the already known facts, opened the door to incompatible hypotheses equally supported by the same observations. For this reason, Mill says, all thinkers with "any degree of sobriety" agree that we should not accept a hypothesis as true simply because it accounts for the known phenomena since "this is a condition sometimes fulfilled tolerably well by two conflicting hypotheses" and probably many more which our minds "are unfitted to conceive." This problem about theory, which has become a mainstay of empiricist philosophy of science, will be considered in detail in the next chapter. But there is another crucial objection to inductivism that should be faced immediately.

Both Bacon and Mill were confident that with sufficient time and observation inductive reasoning would uncover the most basic laws of nature. If we find that bodies always fall at a constant rate of acceleration near the earth independently of their weight, and we have conducted numerous experiments varying every plausible factor we can think of, then it seems we have good reason to suppose that this is a law governing free-falling bodies near the earth. Put simply, the observed instances of free-fall permit us to infer that all instances of free-fall will obey the law, whether observed or not. This is a crucial tenet of inductivist philosophy of science, enshrined by Newton himself in the *Principia* as one of the fundamental "rules for the study of general philosophy:" "Those qualities of bodies . . . that belong to all bodies on which experiments can be made, should be taken as qualities of all bodies everywhere."

But what exactly justifies the inference from how things have been observed nearby to how they would be observed elsewhere? And why does the fact that bodies have fallen at a certain rate in the past give us reason to believe that they will continue to fall this way in the future? The question was first raised by the eighteenth-century Scottish philosopher, David Hume, whose criticism of the argument for intelligent design was mentioned in chapter two. Hume began by pointing out that inductive inferences are certainly not deductively valid. To take Hume's example, the fact that food has always nourished me in the past does not absolutely *guarantee* that it will do so today. It is not impossible that for some reason it fails to nourish me today. Perhaps the inference is instead inductively justified. Since such inferences have been reliable so far, I am right to expect this inference to hold going forward. The problem, Hume pointed out, is that the legitimacy of inductive reasoning is precisely what's at issue. Justifying induction inductively is as hopeless as relying on a hunch to vindicate the power of intuition.

HUME AGAINST MIRACLES

Despite his skepticism about induction, Hume insists "a wise man proportions his belief to the evidence." He means simply that we should believe what is most likely given what we've observed. While this principle might seem obvious, Hume draws some consequences that, at least in his own time, would have seemed outrageous. Consider miracles. Hume notes that most people who believe in miracles do so on the basis of testimony presented in religious texts or reported in the media. But how seriously we take

HUME AGAINST MIRACLES (*cont.*)

testimony for an event should depend on various factors, including how unusual the event is. If a friend tells us he saw a local politician in his neighborhood, we'd probably believe him. If he told us he saw Gandhi, we'd assume he's lying or delusional. In general, we should regard testimony as false if this is more likely than the event testified to. Now when it comes to miracles Hume says it's always more likely that the testimony reporting them is false. For by definition a miracle (rising from the dead, for instance) is a violation of the laws of nature, and the laws of nature have immense empirical support on their side. This argument seems to apply not only to reports from others but even to the testimony of our own senses: if we *see* a person rise from the dead we should not believe our own eyes. For hallucinations and hoaxes do not violate the laws of nature.

Hume's argument has potentially problematic consequences for science as well as religion. For it seems to imply that when a scientist detects something inconsistent with the accepted laws of nature, she should discount the observation as being as improbable as a miracle. Thus, it seems Galileo's observations of lunar mountains were rightly dismissed by Vatican officials, since such irregularities are inconstant with the well-entrenched Aristotelian principle that heavenly bodies are perfectly spherical. But if anomalous observations should always be dismissed as less probable than the established laws of nature, how can science make progress? One way a Humean can handle this problem is by insisting on the replication of experimental results inconsistent with accepted laws of nature. If the results are replicated a sufficient number of times, it may become more likely that the law is false. But notice that this would not be evidence of a miracle, only evidence that the laws of nature are different than we thought. According to Hume's argument, if an event is worth believing in, it is worth changing our laws of nature to accommodate it. But then it looks like evidence of a miracle is not merely unlikely but absolutely impossible.

So why do we trust induction, even bet our lives on it? It may seem obvious that we should eat when we're hungry, because even though it's possible the food will leave us hungry (or worse) this time, natural processes tend to remain constant rather than changing willy-nilly. Thus, Mill noted that the very "ground of induction" assumes the principle that "what happens once will, under a sufficient degree of similarity, happen again, and not only again but as often as the circumstances recur." Similarly, Newton justifies his inductive rule of natural philosophy by observing that "nature is always simple and ever consonant with itself." Both inductivists rely on what we may call the Principle of the Uniformity of Nature (PUN).

So we can invoke PUN to plug the logical gap in our inductive inferences. And if Hume asks what justifies PUN, Mill is ready with an answer: "If we consult the actual course of nature, we find the assumption is warranted." In other words, we can expect PUN to apply at lunchtime because it has applied in the past. But now Hume pounces: this justification of PUN relies on induction, but induction is what PUN was supposed to justify! Hume thinks such circular logic must taint any (non-deductive) justification of induction. Suppose, in order to assuage my new doubts about lunch easing my hunger, I enlist certain physiological laws about the way my body processes the glucose in food and how this affects the neurochemistry in the part of my brain that controls the feeling of hunger. This merely pushes the problem of induction deeper – for why believe that digestion and brain chemistry should continue to work the way they have except that they have done so in the past? As Hume puts the general point: "It is impossible that any argument from experience can prove this resemblance of the past to the future since all these arguments are founded on the supposition of that resemblance."

So did Hume stop relying on induction? Did he cease taking lunch when he was hungry? Certainly not (as portraits of a clearly well-nourished Hume will testify). For he considered induction to be a deeply ingrained human psychological instinct, what he called "custom" or "habit," which we can no more forsake than fear or sexual desire. This does not mean induction is rational, however, any more than our fears or desires are rational. Hume died before Darwin was born, but he probably would have endorsed the notion that induction is an adaptive trait – our ancestors who found it convincing survived better and reproduced more, and that's why it persists in us. Thus he remarks, "Without the influence of custom we should be entirely ignorant of every matter of fact beyond what is immediately present to the memory and senses. We should never know how to adjust means to ends, or to employ our natural powers in the production of any effect." But the practical benefits of induction are not a logical rationale, Hume would be quick to point out, since the fact that it has been useful in the past is no guarantee that it will be useful going forward.

THE NEW RIDDLE OF INDUCTION

Suppose we have solved Hume's problem of induction: from the fact that all emeralds so far observed have been green we are justified in believing all emeralds we observe in the future will be green. In the twentieth century, Harvard philosopher Nelson Goodman put a new spin on the problem. Consider the following term invented by Goodman:

Grue: something is *grue* just in case it was observed in the past to be green or was not so observed and is blue.

Now the same evidence that justifies me in holding that all emeralds are green (all the ones so far observed have been green) seems to justify me in holding that all emeralds are grue (all the ones so far observed have been grue). But I clearly don't believe all emeralds are grue since I don't expect the next one I see will be

THE NEW RIDDLE OF INDUCTION (*cont.*)

blue. Is this prejudice for green over grue justified or merely a kind of Humean habit?

Many object to the gerrymandered grue predicate because it is defined in terms of the more basic green and oddly refers to time. Goodman responds that we can also define green in terms of grue and another predicate:

Bleen: something is bleen just in case it was observed in the past to be blue or was not so observed and is green.

Green: something is green just in case it was observed in the past to be grue or was not so observed and is bleen.

This definition of green, which makes reference to time, may seem odd to us; but for someone raised in the grue/bleen language it might seem odd to treat green and blue as primitive terms. Goodman maintained that we make inductive inferences the way we do because certain predicates are more "entrenched" in our practice, including our scientific practice. But for Goodman this is a contingent fact about our language and practice rather than a matter of logic or objective reason. If "grue" had been entrenched for us, the past grueness of emeralds would lead us to expect them to be grue (hence blue) in the future. But we're stuck with our predicates, just as Hume is stuck with his inductive habits.

Much of modern philosophical thinking about scientific method can be understood as responding to Hume's skepticism (which he applied also to traditional concepts of causation and laws of nature). The great German philosopher Immanuel Kant said he was awoken from his dogmatic slumber by Hume and prompted to invent a "critical" solution to Hume's hyper-empiricist conception of knowledge. Kant's view (which influenced many nineteenth-century philosophers, including Whewell) was that empiricism traded on the misguided assumption that the world could be experienced as it is "in itself" (*an-sich*) free from any involvement of our intellectual faculties. On the contrary, he maintained, the world we attempt to understand in science – the *phenomenal* world – is perceived through a set of hard-wired intellectual categories (time, space, causality, etc.). The world in itself – the *noumenal* world – is forever beyond our grasp. Kant thus offered a synthesis of the empiricist and rationalist conceptions: theory without experience is empty, but experience without theory is blind. While rarely embracing Kant's metaphysical apparatus, the modern conceptions of scientific method that we next consider have also attempted to combine the empiricist and inductivist conviction that science is ultimately grounded on experience with the rationalist and deductivist insight that our intellect contributes in important ways to the construction and evaluation of theories.

Perhaps the most thoroughgoing recent deductivist – and anti-inductivist – scientific methodology is Karl Popper's. Popper accepted the Humean argument that we cannot justifiably derive a universal law or theory from a finite number of empirical observations. He also argued that the empirical success of a hypothesis we have somehow or other devised does not in itself provide reason to believe it is true. We cannot hope to infer the universal truth of our theories from their particular successes since false theories can make true predictions just as well as true ones. If my theory (T) predicts some observation (P), which turns out to be correct, can I infer that my theory is true? This would be tantamount to committing the logical fallacy known as "affirming the consequent." Consider a theory T: "All swans are white." This logically implies the prediction P that a swan observed at a certain time will be white. But it doesn't follow that from a correct prediction the theory is true:

1. T implies P
2. P is true
3. So, T is true (Invalid)

The premise may be true and the conclusion still false for the simple reason that there may be other swans, besides the one we observed, that are not white.

But suppose you observe a black swan at the appointed time. In other words the prediction is false. From this it does follow deductively that the theory is false:

1. T implies P
2. P is not true
3. So, T is false (Valid)

This is a perfectly correct deductive inference. Since the theory says absolutely all swans are white then the observation of even one non-white swan shows it to be false. So Popper's point was very simple: you cannot deductively *verify* a theory by making correct predictions but you can deductively *falsify* a theory by making incorrect predictions. In this way, Popper's deductivism dovetails with his falsificationism.

Even if it is logically improper to deductively infer from correct predictions the *truth* of a theory, perhaps one can at least inductively infer its *probable truth*. But Popper rejected this inference too. You will recall that in his view scientific theories should be highly falsifiable. But all things being equal, the more falsifiable a theory the more improbable it is. If one theory says A will happen while another theory says both A and B will happen, then the second theory is more falsifiable because there are twice as many ways it can be wrong. It will also be less (or not more) probable. This is an unavoidable consequence of the logic of probability (which Popper certainly accepted). Suppose a couple will have two children. I predict the first child will be a boy and leave it at that while you predict that the first will be a boy and the second a girl. I have a fifty percent chance of being right while your chance of success is only twenty-five percent. So Popper concluded: "if our aim is the advancement or growth of knowledge then a high probability (in the sense of the calculus of probability) cannot possibly be our aim as well: *these two aims are incompatible.*"

There are ways of avoiding the consequence that more informative theories must be less probable. One is by focusing on the way probabilities are affected by new evidence. There are perfectly objective ways, which Popper certainly accepted, of calculating how some bit of evidence E increases or decreases the probability of a hypothesis H. For example, I can calculate how much my hopes should be raised by my friend telling me I have won the lottery, if I already know how probable it is that he would tell me this if I hadn't. If I know this is very unlikely, because my friend's not prone to fibbing, his news will significantly increase the probability that I won. The hypothesis that I won is not only informative but also likely. The problem, of course, is that these means of recalculating probability in the light of evidence presuppose but do not explain the "prior" independent probabilities that I assign to the hypothesis (that I won) and the evidence (that he would tell me I won). Without these I can't calculate how the new evidence changes things. But this brings us back to Hume's problem: how do we justify general conclusions, even probabilistically, from limited experience in the first place? In the next section we consider an influential account of how, contrary to Popper, mere predictive success can provide inductive support for a scientific hypothesis.

From a purely logical point of view, the observation that deduction of true predictions cannot prove a theory true seems unexceptionable. Newton was able to predict many of the planetary orbits with great accuracy (as did the Ptolemaic astronomers who came earlier). Yet Newton's theory was not strictly true, and this was revealed by its incorrect prediction of the orbit of Mercury. Einstein however did correctly predict Mercury's orbit and many other surprising phenomena as well. Granted this remarkable empirical success does not prove

Einstein's theory, does it not at least provide some evidence of its truth?

A number of inductive methodologies were developed in the twentieth century to support the intuition that predictive accuracy provides genuine inductive support. Most of these methodologies were developed by philosophers associated with a movement known as *logical empiricism* (or sometimes *logical positivism*). Logical empiricism emerged in Austria between the World Wars among a group of scientists and philosophers known as the Vienna Circle. Many of the Circle's most eminent members, including Rudolph Carnap, Hans Reichenbach and Herbert Feigl, eventually emigrated to the United States and established philosophy of science as a professional discipline (which still thrives). As the moniker suggests, the logical empiricists combined Popper's enthusiasm for the power of modern deductive logic with a strict form of empiricism. While acknowledging the importance of hypotheses, they were sympathetic with the traditional inductivist view that scientific knowledge (unlike metaphysics, which they disparaged as empty) was firmly grounded in immediate experience.

An influential example of this approach to method is Princeton philosopher Carl Hempel's hypothetico-deductive (H-D) model of confirmation. Hempel accepted Whewell's point that scientific progress is guided by hypothetical reasoning, but also retained the inductivist conviction that we learn mainly from experience. His basic model is very simple. The researcher begins with a hypothesis that typically takes a general form, for example Kepler's law that planetary orbits sweep out equal areas in equal times. From this law, together with initial conditions describing the position and velocity of a particular planet at a particular time, a prediction is deductively inferred about the position of the planet at some later time. If the prediction is correct then the law is confirmed; otherwise it is disconfirmed. The more predictions are borne out the better confirmed the hypothesis – and the more reason we have to believe it.

Supplemental rules may be included which give "extra credit" for hypotheses that predict surprising or novel results, are highly precise, or apply to a range of divergent phenomena, and so on. The main point, however, is that correct predictions deduced from a theory are taken to provide inductive evidence for the theory's truth. Hempel concedes that this method is not inductive in the "narrow" sense of allowing derivation of the theory directly from the phenomena. But he insists it is inductive in a "broader" sense "inasmuch as it involves the acceptance of hypotheses on the basis of data that afford no conclusive evidence for it." While far from the Baconian dream of an inductive machine, the H-D model nevertheless conforms to the empiricist image of science as fundamentally guided by direct observation.

It is worth noting briefly two additional features of the H-D model. First, it is not uncommon in science for a fundamental theory to capture more specific laws as "special cases." For example, most of the content of Mendel's laws of inheritance can be derived from more basic principles of molecular genetics. The H-D model suggests why, when this occurs, the more fundamental theory can be confirmed by predictions drawn from more specific laws. Hempel suggests that the confirmation relation is naturally regarded as "transitive" because deduction is transitive: If A implies B and B implies C, then A implies C. Accordingly, if an observation confirms a hypothesis or law, because it is deducible from the law, then it ought also to confirm the broader theory from which the law itself is deduced. For example, Newton's laws (NL) are very general and say nothing specific about the planetary orbits. But Newton showed that Kepler's laws (KL) were logically deducible from his universal laws. So by showing how his theory captures the specific laws of Kepler and Galileo, Newton inherits their inductive support. This makes good sense and the H-D model explains why.

Second, there is an illuminating symmetry between confirmation and explanation in Hempel's view: the predictions derived from KL about planetary positions *confirm* the laws, but by the same token the laws *explain* why the planets are where they are. The same symmetry exists between fundamental theory and derived laws: NL are confirmed by KL and they also explain why KL obtain. This connection between explanation and confirmation provides support to a sort of ampliative inference that is frequently invoked in both science and everyday life: *inference to the best explanation*. Intuitively, the idea is that we ought to believe the hypothesis which best explains the phenomena. The appeal of this inference is not surprising given the H-D model, since confirmation and explanation are the same deductive relationship between hypothesis and phenomenon, viewed from different points of view.

We will return to inference to the best explanation in the next chapter. But it is worth noting how the appeal of this sort of inference might give comfort to the proponents of intelligent design (ID). For their exact position is that ID better explains features of biological systems like "irreducible complexity" than Darwinian evolution does. However, Hempel imposed a strict criterion of adequacy on any explanation: it must give us reason to expect the phenomenon explained. It would be absurd, for example, to "explain" an earthquake by pointing to the day of the week since no given day of the week gives us any special reason to expect an earthquake. Similarly, as I argued in chapter two, the hypothesis of an intelligent designer does not in itself give us reason to expect any particular sort of biological phenomenon, complex or otherwise. Since it fails Hempel's criterion of adequacy, ID receives no inductive support from its alleged explanatory power.

THE PARADOX OF THE RAVENS

Hempel discovered a surprising, and potentially damaging, consequence of his H-D theory. Consider a very simple hypothesis:

(1) **All ravens are black.**

It seems plausible, and it follows from Hempel's theory, that (1) will be confirmed by observations like the following:

(2) "Here is a raven and it's black."

But (1) is logically equivalent to (means the same as) the following:

(3) **All non-black things are not ravens.**

Compare: "All humans are mortal" = "Anything that is immortal is not human."

But if (2) confirms (1), then presumably the following confirms (3):

(4) "Here is a non-black thing and it's not a raven."

However – and here is the paradoxical part – if (4) confirms (3), then it ought also to confirm (1) assuming the following plausible principle, which Hempel accepted:

(*) Logically equivalent statements are confirmed by the same evidence.

So (4) confirms (1) just as much as (2) does. But that means laws like (1) are confirmed by the observation of green shoes, white pigeons, and so on. If only science were so easy!

Hempel's own response is that (4) does confirm (1) but more weakly than (2). He observes that rare implications of a theory naturally confirm it more than commonplace ones. Still, it's hard to believe that I can do ornithology by counting the socks in my dresser drawer.

Beyond inductivism and deductivism

Despite their differences, Popper and Hempel share a commitment to the objectivity of scientific reasoning. They both conceive of scientific method as a form of argument involving deduction of statements from hypotheses followed by comparison of those statements with direct experience. The objectivity of science is therefore secured by the absolute status of logic and by observations that do not presuppose the truth of the theory under examination. It is a matter of simple, deductive logic whether a theory predicts some phenomenon; and it is a matter of public, empirical fact whether the phenomenon occurs.

But in the second half of the twentieth century both of these assumptions were called into question, and along with them the objectivity of science itself. A leader in this assault on the traditional, logically oriented approach to scientific method was Thomas Kuhn – probably the most influential philosopher of science since Mill. Kuhn was trained as a physicist but became intrigued by the history of his field. He found that this history did not conform to the legends passed down in physics classes nor to the "rational reconstructions" proposed by philosophers. Scientists tended to view history from the perspective of the victors: a series of noble struggles on the inevitable road to contemporary theory. Philosophers favored excessive logical abstraction, convinced of the essential rationality of science. But if we take history seriously, Kuhn maintained, then scientific progress and rationality are very different from these idealized conceptions.

Central to Kuhn's own model of scientific development, presented in his groundbreaking work *Structure of Scientific Revolutions*, is the notion of a "paradigm." A paradigm is a major theoretical achievement that establishes an "exemplar" or framework for future research in a given field. Young scientists are inculcated in the assumptions and methods of the paradigm and commit themselves to its articulation and application. Much of everyday or "normal science" is devoted to solving theoretical and empirical "puzzles" that the paradigm presents. For example, in the paradigm of physics established by Newton's *Principia*, scientists developed an elegant mathematical formulation of his theory, applied the laws to gases and fluids, and resolved anomalies in the planetary orbits.

Contrary to Popper, in Kuhn's view scientists are decidedly not interested in falsifying their paradigm because without a paradigm there is no systematic inquiry at all, only unguided fact-collecting and philosophical speculation. So for Kuhn, the difference between astronomy and astrology, and between science and non-science generally, has little to do with falsification. Astronomy, as in the paradigm of Copernicus, provides clear guidelines and techniques for puzzle-solving; but astrology offers no such systematic means for dealing with predictive or explanatory failures. Two different astrologers may have entirely different diagnoses of the failure, and no principled way of deciding between them.

History also shows that the puzzle-solving effectiveness of paradigms inevitably wanes when intractable anomalies emerge as the paradigm is extended with increasing precision into new domains. Eventually the mounting failures of the paradigm lead to a state of "crisis" and finally to a revolution and the ascension of a new paradigm. One of the remarkable features of a scientific revolution is that, as in a political revolution, there is a breakdown of rational argument between proponents of the old and new regimes. The reason, according to Kuhn, is that the paradigm itself determines the proper methods for doing science. For a paradigm is not only a theoretical exemplar, but also a "disciplinary matrix" that fixes the values and aims of scientific inquiry. There is no question of convincing the other side that your paradigm is better since you don't agree what constitutes good science in the first place. "In the partially circular arguments that result," Kuhn explains, "each paradigm will be shown to satisfy more or less the criteria that it sets for itself and to fall short of a few of those dictated by its opponents."

So a revolution succeeds not because the old paradigm is falsified nor because the new one is better confirmed by the rules of an abstract inductive logic. Rather the new paradigm offers the next generation of scientists an interesting set of puzzles together with a promising set of techniques for unraveling them. The members of the older generation are not won over – they either fall in line or are relegated to the margins of mainstream science. As one of the pioneers of quantum theory, Max Planck, put it, "a new scientific truth does not triumph by convincing its opponents and making them see the light, but rather because its opponents eventually die, and a new generation grows up that is familiar with it."

So, for Kuhn, there is no paradigm-transcendent logic of science, whether inductive or deductive. He also rejected the other basis for scientific objectivity shared by Popper and Hempel, that observations in science can be characterized in a way that does not presuppose the theory under consideration. Drawing on the work of the philosopher N. R. Hanson, and the field of gestalt psychology, Kuhn argued that it is futile to appeal to independent observations to adjudicate competing paradigms since the conceptual structure of the paradigm determines the content of the scientist's perceptions. Just as the figure opposite (Figure 2) may be perceived as a duck or a bunny, depending on orientation and previous conditioning, different scientists will perceive the same objects differently according to the orientation of their paradigm.

Where an ancient astronomer sees the sun rise, a modern sees the earth turn; when an Aristotelian looks at a pendulum she sees something diverted from its tendency to natural rest, while a Newtonian sees something nearly achieving inertial motion; where Priestley sees "dephlogisticated air," Lavoisier sees oxygen; and so on. In a sense, these shifts are even more dramatic in science than in gestalt experiments. With some effort we might come to see the duck and rabbit as alternating effects

Figure 2 The infamous duck-rabbit

of an underlying set of pencil lines that themselves are seen as neither duck nor rabbit. But there seems to be no analogous recourse to the underlying data in scientific observation, since our best, most fundamental way of characterizing the phenomena is given by science itself. "Swinging stone aiming at natural rest" is not for the Newtonian a different take on some underlying phenomenon, but a confused and mistaken characterization of the approximately inertial motion of a pendulum.

From considerations such as these about the absence of any theory-neutral scientific observation, Kuhn was led to the rather startling conclusion that "we may want to say that after a revolution scientists work in a different world." We will consider the radical implications of such a conclusion in the next chapter. For now it is enough to appreciate that the "theory-ladenness" of observation poses a serious threat to the logical empiricist approach to scientific methodology. If there is no theory-neutral observation, then the empirical world can no longer serve as an objective arbiter among competing hypotheses.

Although Kuhn thinks methodologies are paradigm-relative, he certainly does not advocate competing methods *within* normal science. On the contrary, dogmatic adherence to the rules of a paradigm is the key to progress in normal science. Another unorthodox philosopher from the post-war period, Paul Feyerabend, advocated wholesale methodological anarchism. Something of a philosophical provocateur, Feyerabend had studied theatre as a young man and his lectures are said to have sometimes approximated performance art. His major work, *Against Method*, opens with the promise to establish that "given any rule, however 'fundamental' or 'rational' there are always circumstances when it is advisable not only to ignore the rule, but to adopt its opposite." Feyeraband's favorite historical case study is the Copernican Revolution, and the scientific practice of Galileo in particular. In order to advance his Copernican agenda, Galileo happily employed induction right alongside "counter-induction" (inferences intentionally contrary to clear evidence), ad hoc hypotheses, appeals to authority, "propaganda," and even deception. Feyerabend regarded such methodological opportunism as perfectly appropriate and indeed essential for scientific progress, then and now.

Feyerabend was fond of slogans such as "anything goes" and "let a hundred flowers bloom," convinced that a constant proliferation of methods and perspectives contrary to the accepted canons of reason was necessary to avoid stagnation and dogmatism: "Ideas which today form the very basis of science exist only because there were such things as prejudice, conceit, passion; because these things opposed reason; and because they were permitted to have their way." For Feyerabend, this meant taking seriously even styles of thinking entirely alien to rationalistic Western science, such as Chinese medicine, Voodoo and Azande witchcraft. While Feyerabend's reading of history has not gone unchallenged, his attack on methodological universalism encouraged more pluralistic approaches within the philosophy of science, which are now common, as well as post-modern and social-constructionist critiques of science (about which more in the next chapter).

Recall from chapter two that one of the most serious objections to the falsificationist demarcation criterion is that predictions are not derived from our theories in isolation, but only in conjunction with numerous empirical, experimental and mathematical assumptions. For example, the prediction of the return of Halley's comet from Newton's laws required various assumptions about the positions and masses of other bodies, the accuracy of the telescopes and other instruments used to collect this data, the reliability of the mathematical calculations used in the derivation, and so on. So, had the prediction failed, scientists would have faced a choice of where to lay the blame, and neither deductive nor inductive logic seems to legislate a choice. The point was stressed at the turn of the last century by the French philosopher Pierre Duhem, and later made a cornerstone of the "holistic" approach to scientific knowledge developed by the influential American philosopher W. V. O. Quine. Quine suggested our knowledge of the world is closer to a "web of belief" than a deductive or axiomatic system. When there is a problem – an internal inconsistency or a conflict with experience – then adjustments must be made. But nothing logically requires that adjustments be made in one part of the web rather than another: "any statement can be held true come what may, if we make drastic enough adjustments elsewhere in the system."

If Quine is right, then it would seem that the philosophical goal of reducing the scientific method to absolute rules of logic or pure reason is misguided. But where does this leave the philosophy of science in its effort to understand scientific knowledge? Quine maintained that the theory of knowledge generally (epistemology) ought to be "naturalized." Since human knowledge is at bottom a psychological phenomenon, we should investigate it using the tools of empirical psychology:

"Epistemology, or something like it, simply falls into place as a chapter of psychology and hence of natural science. It studies a natural phenomenon, viz., a physical human subject." Taken to its logical limit naturalism should apply even to philosophy. Indeed, one of the more interesting developments in recent philosophy has been the emergence – some would say *resurgence* given the close historical association between philosophy and science – of the empirical study of philosophy itself, a movement known as "experimental philosophy."

As distinctive as Popper, Hempel, and Kuhn are, they all seem to endorse a distinction, drawn most clearly by the logical empiricist philosopher Hans Reichenbach, between the "context of discovery" and "the context of justification." Essentially, Reichenbach maintained that epistemology should be concerned with the logical analysis or "rational reconstruction" of science (the context of justification) not its psychological or sociological causes (the context of discovery). Popper and Hempel fully accepted this conception of the philosophy of science. And even Kuhn, although he had much more to say than the other two about the process of discovery in historical context, seemed to despair of finding the ultimate nature of scientific discovery: "What the nature of that final stage is – how an individual invents (or finds he has invented) a new way of giving order to data all assembled – must here remain inscrutable and may be permanently so." Naturalized approaches to philosophy of science exactly reverse this hallowed distinction, putting the actual process of discovery prior to the logic of justification and urging a thoroughgoing historical and empirical study of this process. As Ronald Giere has encapsulated the naturalist manifesto: "the study of science must itself be a science."

Unsurprisingly, naturalism has been met with resistance. Some detect a vicious circularity in the naturalist program: using science to study science presupposes the very thing you are trying to understand. You can't learn about your own eyeballs by looking at them. Naturalists have replied that we can certainly use one science to examine another (using psychology to study physics, for example) just as we can hire an ophthalmologist to examine our eyes. In fact, it may be instructive to turn psychology on itself and ask what cognitive processes are involved in the science of cognition. Sociology has also been widely promoted as the ideal "science of science," as we'll see in chapter five, but many naturalists have been content with a more pluralistic approach.

More problematic is the notion that we can use science not merely to understand it as a process but to validate or justify the knowledge it produces. This does seem hopelessly circular, like using the Bible to prove that God exists or literally pulling yourself up by your bootstraps. This raises a second charge against naturalism, one leveled so routinely that it carries its own (dis)honorific label, the "naturalistic fallacy." In this context, the fallacy would involve inferring the *truth* of science from what we learn scientifically about the *causal processes* governing the practice of science. But these are very different matters. As the defenders of the discovery/justification distinction insisted, it seems I can tell you all about how a person comes to believe what they do, including a scientist in the course of inquiry, without having any idea whether their beliefs are true. And vice versa. We will see in chapter five that this distinction between the causes vs. reasons for scientific belief is not so clear-cut when we examine science in a social context. Nevertheless it seems that the empirical study of science must leave open a fundamental philosophical question about science: has it achieved its aims? This is the topic of the next chapter.

4

The aims of science

We have so far discussed the origins, essence, and methods of science. We now consider its aims. Surprisingly (or perhaps by now you won't find it so surprising!) there is little agreement among philosophers about what science is for or what it has achieved. Nevertheless, certain views about the purpose of science that were once common are no longer widely subscribed to. For the ancient Greek philosopher Epicurus, the aim of science was to help dissolve groundless fears about death and the unknown: "If our suspicions about heavenly phenomena and about death did not trouble us at all . . . then we would have no need of natural science." For many medieval and early modern philosophers, such as Robert Boyle, science was primarily for the glory of God: "the knowledge of the works of God proportions our admiration of them, they participating and disclosing so much of the inexhausted perfections of their Author." But even these attitudes seem to presuppose an aim for science that is still prevalent, namely to uncover the truth. For unless science tells us how things really are, then how can it relieve superstition or glorify God's workmanship? But whether or not happiness or piety are aided by science it is reasonable to ask whether truth is what we should expect from science. Perhaps it is enough, as Bellarmine suggested to Galileo, for science to "save the appearances." Or perhaps science should simply serve the public interest – "improve man's estate" as Bacon put it. In this chapter we evaluate the most prominent conceptions of the ultimate aims of science.

Scientific realism

The modern sciences present us with an exotic and unfamiliar world. The universe is an immense four (or more) dimensional "space-time manifold" populated by billions of galaxies consisting of stars, dust, occasional black holes and plentiful "dark matter." The stars, planets and satellites are composed of numerous elements that are in turn composed of elementary particles that behave in somewhat predictable ways under the influence of four fundamental forces. The overall entropy or disorganization of the universe as a whole is increasing, moving slowly but inexorably toward a "heat death" in which even the atoms disintegrate. Despite this death march, the elements are arranged in such a way as to allow for the emergence of complex biological systems that reproduce themselves and evolve over time. Among the millions of life forms on Earth, one has produced highly sophisticated social structures and cultural artifacts, including language, technology, and science. The science of the early twenty-first century declares the universe as a whole began with a massive "big bang" about twelve billion years ago and will either end with an inverse "crunch" or go on expanding forever.

In developing this picture of reality over the last 2500 years, it is hard to deny that modern science has been extremely successful in two respects. First, it predicts with remarkable accuracy the observable behavior of most physical, chemical, and biological systems. This predictive accuracy tends to vary inversely with the complexity of the systems involved, of course. But even extremely complex economic and weather systems can now be anticipated within large margins of error. Related to this predictive or empirical success is the immense technological achievement of science. Modern medicine, especially drug therapies and diagnostic technologies, owe their success directly to advances in chemical and electronic sciences. Physical science has a long history of "successful" application to military technologies, from the computation of projectile motion in the sixteenth century to the cruise missiles of today. Current research in the physics of tiny "nano-systems" will likely produce yet another revolution in computing and possibly energy technology. Food and clothing, at least for most of us, are produced using technologies derived from various molecular sciences. Humans now permanently inhabit space and, as I write, the robotic spacecraft Cassini is transmitting data from near the surface of the Saturnian moon Enceladus.

But for all this success, should we presume that the exotic account of our universe summarized above captures the ways things really are? *Scientific realism* is the claim that modern scientific theories provide a true (or approximately true) account of the world. *Anti-realism* comes in a variety of forms, as we will see, but the essential view is that the aim of scientific theories is not to provide a true account the world. The most compelling argument for realism is based on one or both of the two kinds of successes just described. The argument is that the best, and perhaps only, explanation for the immense empirical and technological success of modern science is that it is more or less true. For example, the best explanation for the remarkable success of the standard model of the atom is that there are electrons, protons and neutrons arranged in roughly the way the model says. Since defenders of this argument sometimes say the success of modern science would be an incredible coincidence or even a "miracle" unless science was basically true, I will refer to this as the "no miracles" argument for scientific realism.

> **No Miracles Argument:** Realism is the best (or only) explan-
> ation for the empirical and technological success of modern
> science.

Realists observe that the pattern of reasoning used in this argument, "inference to the best explanation" (IBE), is commonly employed in both science and everyday life. For example, although I have never asked them directly, I believe that my next-door neighbors are married because this best explains why they live together and wear matching gold bands. Although it is not directly observable, astrophysicists believe in "dark matter" because its existence explains various known gravitational effects such as the rotation of galaxies. By the same reasoning, the scientific realist believes in the truth or near truth of modern science because this best explains its success. It is possible, of course, that science is successful even though the world is nothing like its theories say, just as it is possible that my neighbors are merely cohabiting and happen to share a taste in rings. But since true theories seem more likely to be successful than false ones – though we will see below that serious doubts can be raised on this point – and since matching gold bands are not normally worn by couples living together, these possibilities are not as reasonable to believe.

At least one opponent of scientific realism, the philosopher Arthur Fine, has charged that this argument for realism merely "begs the question." In other words, it assumes exactly what it purports to prove. In particular, the no miracles argument relies on IBE even though the legitimacy of such reasoning is exactly at stake in the realism debate. Anti-realists don't accept IBE when it is applied to particular scientific theories. For example, they don't consider the explanatory power of quantum theory sufficient evidence for its truth. So it's hardly fair dealing to use the very same form of reasoning in support of a philosophical theory like realism. Fine writes: "To use explanatory success to ground belief in realism, as the explanationist defense [the no miracles argument] does, is to employ the very type of argument whose cogency is the question under discussion. In this light the explanationist defense seems a paradigm case of begging the question."

While this objection has been widely applauded by anti-realists, its real impact is weak. In the strict sense, an argument begs the question when it assumes as a premise its own conclusion. The arguer thus merely "begs" or pleads the question be resolved in his favor rather than providing reasons. Example: The death penalty is wrong because execution is an unethical form of punishment. Clearly, the no miracles argument does not beg the question in this sense, since IBE (the premise) is not the same doctrine as scientific realism (the conclusion). According to a slightly weaker standard, an arguer begs the question if she relies on a premise that no one could have reason to accept unless they already accepted her conclusion. This doesn't apply either. One could certainly have reasons to accept IBE that have nothing to do with scientific realism. Finally, it is sometimes said that an arguer begs the question if she employs a premise she is unwilling or unable to defend. But the realist is prepared to defend IBE, by citing its success in other contexts or in everyday life. The anti-realist may not be persuaded, of course, but that doesn't mean their opponent's reasoning is fallacious. Indeed, to simply insist that the realist's premise begs the question amounts to committing what the nineteenth-century logician Augustus DeMorgan called the "opponent fallacy:" "it is the habit of many to treat an advanced proposition as begging the question the moment they see that, if established, it would establish the question."

The no miracles argument doesn't beg the question; but there are other more serious objections it must face. One is that there are alternative ways of explaining the success of science that don't force a choice between truth and miracles. For example, the eminent anti-realist Bas Van Fraassen offers an evolutionary explanation: "I claim that the success of current scientific theories is no miracle. It is not even surprising to the scientific (Darwinist) mind. For any scientific theory is born into a life of fierce competition, a jungle red in tooth and claw. Only the successful theories survive – the ones which in fact latched on to actual regularities in nature." This objection is clever, but misleading. If by "survival" of a theory we mean its acceptance by scientists, this can't explain its success. For it survives in this sense because it's judged more successful by scientists: the success explains the survival, not vice versa. Analogously, the reason a species (or organism) survives so well over time is because it's more successful (at finding food, reproducing, etc.) than its competitors, not vice versa. To explain the success of the species we need to cite the traits that make it good at finding food and mates. The trait that explains the success of modern science, and so the survival of its theories, the realist suggests, is truth. Furthermore, modern scientific theories are often successful in "novel" ways that were not part of the reason for their original formulation or acceptance. For example, Einstein's special theory of relativity accurately predicts the time-dilation (slowing of time) experienced by atomic clocks on high-speed jets, though the theory was formulated and accepted long before such jets (or clocks) existed. This is analogous to a species succeeding despite a change in the environment in which it struggled for existence. Such success cannot be explained by the reasons for its earlier survival.

Although these initial objections to the realist argument are not conclusive, the anti-realist is armed with two arguments of his own, each as intuitive and powerful as the no miracles argument. The first makes a "pessimistic induction" from the history of science to undermine the realist inference from empirical success to truth. The history of every science is littered with theories that were empirically successful but which we now consider quite wrong. For example, the phlogiston theory of eighteenth-century chemistry enjoyed significant success accounting for various processes like combustion and calcification even though the theoretical entity it relied on was repudiated by modern chemistry since the discovery of oxygen. Many discarded theories were successful in their day, including Ptolemy's astronomy (with its geocentric structure and concentric spheres), the caloric theory of heat, Descartes' vortex theory of the orbits, the electromagnetic ether, J. J. Thomson's "plum pudding" model of the atom, and so on. In fact, it seems that nearly every major theory was eventually disproved, even Newton's which dominated science for 200 years. Hence, if we take history seriously the fair inductive inference, pessimistic as it is, ought to be that our current theories will also be rejected in the long run. At a minimum, the pessimistic induction undercuts the realist assertion that empirical truth is needed to account for success. As the philosopher Larry Laudan puts it: "The inescapable conclusion is that insofar as many realists are concerned with explaining how science works, and with assessing the adequacy of their epistemology by that standard, they have thus far failed to explain very much."

There is a second serious objection to realism. Recall that Mill raised a worry about Whewell's endorsement of hypothetical conjectures that go beyond the available evidence. His worry was that since the evidence for such conjectures "underdetermines" the hypothesis, in the sense the evidence does not point to one unique hypothesis, one could have two (or more) distinct hypotheses supported by the same evidence. But if evidence provides reasons for belief, as the realist maintains, it could then provide reason for belief in multiple incompatible hypotheses. Whewell replied that underdetermination simply does not occur in the course of actual science and can therefore be safely ignored. But this is not so obvious. Newton's theory posited absolute notions of space and time and therefore absolute velocity, but it can be shown that all the same predictions follow from a version of Newton's theory using only a primitive notion of absolute acceleration and no absolute space or time; Lorentz's "absolutist" alternative to Einstein's theory of relativity seems to imply all the same observations while maintaining a privileged reference frame, contrary to Einstein; and the various distinct "solutions" of string theory mentioned in chapter two seem to be empirically equivalent. In any case, it seems to be sufficient for Mill's point that there exist merely *possible* alternative conjectures which, as he put it, "for want of anything analogous in our experience, our minds are unfitted to conceive." In other words, the real problem is that realism engenders a kind of epistemological promiscuity licensing the potential embrace of inconsistent beliefs.

This is the lesson of underdetermination drawn by many contemporary anti-realists. The empirical success of a theory, they maintain, cannot require belief in its truth since we can envision another theory which makes all the same predictions but which makes different, incompatible claims about the unobservable world. Rather than believing our hypotheses, we should merely hold that they "save the appearances." Along with the historical examples given above, the basic problem of underdetermination can be grasped by considering the common practice of fitting data points to curves plotted on a graph. Suppose we are interested in the relationship between two variables and have recorded numerous data points on a graph with one dimension for each variable. For any finite set of data there will be an infinite number of curves that can be fitted to those points (try it yourself). Only one of these curves can describe the real relationship between the variables – but the data doesn't tell us which.

We may prefer to work with paths that strike us as simple or "elegant" of course. This is the upshot of the famous principle "Ockham's razor:" "do not multiply entities beyond necessity." In this case, why prefer a circuitous path connecting the points when a more direct one covers the same data? Ockham's razor clearly makes good practical sense; but does it provide evidence of truth? How can we know in advance that nature is simple? Moreover, what do we do when the underdetermined hypotheses are equally simple, like two zig-zag paths that cover the same points but are mirror opposites? Again, in most realistic cases comparative simplicity is not easy to judge. Is Newton's theory simpler than Einstein's, since there is only one relation of simultaneity in absolute time, or is Einstein's simpler because it dispenses with absolute time altogether? For the Aristotelian, circular motion seems simplest because its beginning and end are one, while for the Cartesian, rectilinear motion seems simplest because it does not involve change of direction. Theoretical elegance may be largely in the eye of the beholder.

In order to evaluate scientific anti-realism it is useful to view it as a form of skepticism. Skepticism is the ancient philosophical doctrine that we know very little. Scientific anti-realism can be understood as a moderate form of skepticism since it claims that science doesn't provide us with knowledge beyond the observable. In fact, the anti-realist arguments just discussed are quite similar to two classical skeptical arguments, both of which Descartes employed in his quest for new foundations for knowledge. In the first of his *Meditations on First Philosophy,* Descartes observes about the senses: "it is prudent never to trust completely those who have deceived us even once." This is analogous to the pessimistic induction, which emphasizes that the scientific method has "deceived" us in the past, by producing false but empirically successful theories, and so cannot be "trusted" in the present. Now Descartes himself rejects this reason for skepticism about the senses. Although he was deceived from a distance or when the light was poor, it does not follow that he is deceived now that the conditions are better. A similar reply to the pessimistic induction is possible if the scientific realist can find a relevant difference between the conditions supporting past theories like phlogiston vs. modern theories like oxygen. The realist might point out, for example, that oxygen theory has been successful in more dramatic ways and over a more sustained period than phlogiston theory.

The problem is that the criteria that seem to favor the current theory might in the long run turn out to have misled us. For we know in retrospect that enthusiasm for the truth of past theories was misplaced. This is the gist of the next skeptical argument Descartes employed against the senses, the famous "dream hypothesis:" "I see plainly that there are never any sure signs by means of which being awake can be distinguished from being asleep." This is analogous to the underdetermination argument: just as our sense experience is compatible with either waking or dreaming so that we can't say for sure which is really the case, our empirical data is compatible with numerous different theoretical explanations. So we should admit that we really don't know.

However, there is another point Descartes makes about dreaming, seemingly as an afterthought near the very end of the *Meditations*. Contrary to his earlier assertion, he notes there is a "vast difference" between dreams and waking: "dreams are never linked by memory with all the other actions of life as waking experiences are." In other words, waking experience forms a coherent, interconnected whole while dreams are fragmented both within and among themselves. The realist might assert an analogous difference between accepted scientific theories and discarded or underdetermined alternatives: current theory is related in numerous intricate ways to other theories and sciences. For example, quantum theory is directly involved in research in particle physics, chemistry, optics, and cosmology. And the curves scientists choose to plot on graphs, among all those possible, will naturally reflect connections to other areas of research. For example, if the relationship between wavelength and frequency for microwave radiation is already well-established, then from the existing data on gamma radiation we will infer a relationship that conforms to the established pattern. But discarded past theories are dead-ends, and as yet unconceived alternatives obviously have no real connections to ongoing research. It is true that past theories did, and unknown alternatives might someday, have interesting connections to other theories in their own times and domains. Nevertheless, the more unification and integration is found among the modern sciences, the less likely it seems it will have all been a dream.

SCIENTIFIC REALISM AND THEISM: STRANGE BEDFELLOWS?

I have suggested an affinity between skepticism and anti-realism. Is there a corresponding allegiance between religious belief and scientific realism? Historically it would seem not, since some of the most prominent defenders of scientific anti-realism have also been theists: Cardinal Bellarmine (who said it is enough for astronomy to "save the appearances"), Pierre Duhem (who said empirical observation "does not have the power to transform a physical hypothesis into an indisputable truth") and Bas Van Fraassen (who said the aim of science is not truth but only "empirical adequacy") are all theists. Yet in a chapter of his modern classic *The Scientific Image* called "Gentle Polemics," Van Fraassen suggests that the scientific realist's arguments parallel St. Thomas Aquinas' famous "five ways" of proving the existence of God. For example, the "first way" is that things in motion are moved by other things, but since this chain of movers cannot go back infinitely there must be a first "unmoved mover." This being, Aquinas says, "everyone understands to be God." Analogously, Van Fraassen suggests, the realist argues that regularity in the phenomena need to be explained by something else, but since the chain of "explainers" can't be infinite, we must "arrive at something which explains, but is not itself a regularity in the natural phenomena." This stopping point in explanation, the realist's unmoved mover, is the world of unobservable entities and processes posited in modern physics.

But, as Van Fraassen notes, the realist's logic is subject to the same sort of objection that is commonly raised against Aquinas' first way: if the regress of movers can end with God why can't it just as well end with the world? Why can't the big bang be the unmoved mover? Analogously, if the demand for explanation can come to an end at the level of unobservables, why can't it come to

**SCIENTIFIC REALISM AND THEISM:
STRANGE BEDFELLOWS? (cont.)**

an end before that at the level of observed regularities? Why isn't it enough for science simply to "save the phenomena" as Bellarmine and Duhem suggest?

This raises an important question about the realist view of science: does the aim of explaining observable phenomena in terms of underlying causes go "all the way down" with no end? Suppose one of the basic string theory models is confirmed. It seems like the question will then arise: what explains why strings vibrate in just the way they do? Perhaps there is no end to the depth of scientific explanation. Aquinas argues that the chain of movers cannot be infinite because unless there is a first mover, like the hand that cranks a set of gears, none of the gears would receive motion. This worry does not seem to apply to explanation. Even if there is no "ultimate" or deepest level of explanation, this does not undermine the explanations provided at the higher levels. So granted the superficial analogy between the theologian's first way and the realist's "no miracles" argument, it's not clear that they must both arrive at the same inexplicable explainer.

Scientific realism has been forced to adapt in the face of the powerful anti-realist critique. One way of adapting is to rely on a modified version of the no miracles argument that focuses on progress rather than truth. Instead of inferring the literal truth (or even approximate truth) of our current theories from their success, the realist can infer increasing *truthlikeness* as the best explanation for the increasing empirical or technological success exhibited in the overall history of a given science. For example, the best explanation for the increasing empirical and technological success of early twentieth-century atomic models from Thomson through Rutherford, Bohr, and beyond, is that these models are increasingly similar to the world (closer to the truth). And the increasing success of astronomy from Ptolemy through Galileo, Newton, and Einstein is best explained by real advances in our knowledge, even if the final truth is still a long way off. This brand of realism, which we might call *progress-realism* in contrast with the more familiar "truth-realism" is aptly summarized by Popper: "while we can never have sufficiently good arguments in the empirical sciences for claiming that we have actually reached the truth, we can have strong and reasonably good arguments for claiming that we have made progress towards the truth."

Progress-realism is not as vulnerable as truth-realism to the anti-realist assault. The pessimistic induction undermines the inference from success to truth by citing numerous historical cases of successful but false theories. In order to deflect in the same way the progress-realist inference, an anti-realist opponent would need to document historical sequences of increasingly successful sciences that nevertheless brought no genuine theoretical progress. If there are such histories, they are much more rare than the various successful but false theories from the past. So there is a much weaker inductive basis for pessimism about the progress of modern science.

Progress-realism is also less vulnerable to the underdetermination argument. As we have seen, it is relatively easy to identify or concoct empirically equivalent but theoretically distinct versions of particular theories. It is less easy to do so for whole sequences of increasingly successful theories. Consider again the sequence of atomic models developed in the early twentieth century. We might be able to construct or conceive a model that makes the same predictions as Thomson's (1904) "plum-pudding" model of the atom. But to underdetermine progress-realism, we would need to also provide alternative empirically equivalent versions of Rutherford's (1911) "Saturnian" model, Bohr's (1913) quantized version of the Saturnian model, and so on. Moreover, since the progress-realist is claiming that the empirical progress displayed by the sequence is best explained by genuine theoretical progress, the anti-realist would need to show that the theoretical transitions within the alternative sequence can account just as well for the empirical improvements we know obtained. For example, one of the main empirical problems with Rutherford's atom was that it predicted a continuous spectrum of radiation from all the possible paths of the orbiting electrons. But experiment showed only certain "signature" frequencies depending on the kind of atom (e.g. hydrogen). Bohr's atom predicted these "spectral lines" because it allowed only certain orbital paths with discontinuous "quantum leaps" between them. The anti-realist would therefore need to provide not only empirically equivalent versions of the Rutherford and Bohr atoms, she would also need to show that these alternatives are different from one another in a respect that would account for the characteristic spectral lines. Put simply, it is a much more difficult task to show that the dynamic improvements between theories is underdetermined than merely showing that the success of isolated theories is underdetermined.

Another new version of realism, somewhat related to progress-realism, is known as *structural realism*. This is the view that modern science achieves a true or "truer" account of the world only with respect to its mathematical structure rather than its intrinsic qualities or nature. Like progress-realism, structural realism attempts to block the pessimistic induction. The structural realist points out that the most successful past theories are usually not abandoned entirely; rather their mathematical structure is carried over into the theory that replaces it. For example, although the fluid ether of Fresnel's optical theory is rejected as a medium for light transmission in Maxwell's theory of the electromagnetic field, Maxwell's equations preserve a mathematical formalism quite similar to Fresnel's. Light isn't wave

motion in the ether, but such motion is similar in form to
Maxwell's field model. By focusing on structure rather than
mechanism, form rather than content, structural realism hopes to
have it both ways. On the one hand, it can still rely on a kind
of no miracles argument: science is increasingly successful
because it latches onto the real (mathematical) structure of the
world. On the other hand, structural realism avoids the
pessimistic induction: the most important parts of successful past
theories are preserved in current theory.

Despite these important virtues, structural realism faces
serious problems as well. First, structural realists need to show
that it is the mathematical aspects of theories, not their content,
that accounts for their success. This might be somewhat plausible for some highly mathematical theories in physics; but it is
harder to see how successful theories in areas like biology,
geology, and psychology could be stripped of their content and
reduced to mathematical formulas. Second, it is not clear that
the structure and content of theories can be as neatly separated
as the structuralist view implies. The mathematical quantities
used in modern space-time theory, for example, such as continuous affine transformations, are successful precisely because they
apply to a world with exactly isomorphic qualities: dense,
connected, four-dimensional, etc. In such cases, if we state the
laws in their full mathematical dress it seems we have specified
all the content the theory has. In these cases, structural realism
collapses into progress-realism. Finally, to the extent we can
abstract mathematical structure entirely from content or natures,
it doesn't seem obvious that what remains should be considered
science. Such structures, like the 11-dimensional spaces of string
theory, may have immense formal interest to the mathematician,
as may structures abstracted from biology, music, or linguistics.
But merely as structures they do not represent the physical
world at all. And for the realist that is the whole point of science
after all.

Varieties of anti-realism

If theoretical claims about electrons, dark matter, cosmic rays, etc., should not be taken as true or even "truthlike" descriptions of the world, then what exactly is their function? *Instrumentalism* is a strand of anti-realism, popular in the early twentieth century, which holds that theories are best understood as tools or instruments for organizing experience rather than straightforward claims about the world. In Duhem's terms, they are devices for "saving the phenomena." So, for example, when a particle physicist asserts that a cloud chamber photograph shows the path of an electron, he is really asserting the value of electron theory to account for this and similar phenomena. Instrumentalism was supported by traditional empiricists who wanted to avoid the admission of hypothetical entities, as well as by American pragmatists like John Dewey who analyzed concepts in terms of utility. Closely associated with instrumentalism was the effort to construct "operationalist" definitions of theoretical terms that refer only to specific experimental procedures. Thus "temperature" could refer not to the unobservable motions of molecules but rather to various measurements involving the expansion of mercury, and "intelligence" could refer not to some intrinsic mental capacity but rather to quantifiable performances on standardized tests.

The main difficulty with instrumentalism is its implausible account of the meaning of theoretical claims and concepts. Most practicing scientists take their assertions about theoretical entities as straightforward descriptions (right or wrong) about the world, not indirect praise for the usefulness of certain ideas. Indeed, they would mostly say that some descriptions are more useful than others *because* they are more accurate, not the other way around. The strangeness of interpreting theories as mere tools for organizing present experience is brought out clearly in sciences like cosmology and paleontology, which largely concern events in the remote past or future. Indeed it seems to be inherent to the concept of the big bang, and perhaps of dinosaurs, that they existed prior to any human experiences. It is true that big bang theory helps make sense of current experiences, such as galactic red-shift and cosmic background radiation, but again that's because it refers to a remote and hypothetical cause of these of experiences.

According to instrumentalism, it makes no more sense to ask whether string theory is true than to ask whether Spanish or Penicillin is true: these are all tools for various ends, not claims in themselves. Claims on behalf of the theory are therefore construed as claims for its effectiveness. A similar version of anti-realism, *semantic reductionism*, holds that theories are indeed claims, but disguised claims about experience rather than about unobservable entities. In the reductionist view, to say that there is a black hole at a certain location in space is to specify what sorts of experiences (especially astronomical observations) one would expect under various conditions. You will recall that classical empiricists, such as Mill, eschewed hypotheses. The semantic reductionist allows hypotheses but employs various logical techniques to show that their content or meaning is fundamentally empirical.

An interesting and influential example of such reductionism from mid-twentieth-century psychology is the *behaviorism* of B. F. Skinner and his followers. Influenced by classical and logical empiricism, the behaviorists were dubious of speculative psychological theories like Freud's, which posited drives and forces unobservable to the researcher (and to the patients themselves). The behaviorists therefore attempted to reduce all seemingly subjective and internal psychological terms such as "belief" and "desire" to objective and observable behavior. To say "Geoff believes it's raining," for example, might be interpreted as a short-hand way of saying "Geoff will take his umbrella to the office;" "Geoff will say it's raining if asked," and so on.

Despite a great deal of technical work aimed at reducing theory to observation, semantic reductionism, like instrumentalism, failed to provide a workable conception of theoretical language. One problem is that theoretical concepts typically have a richness that allows them to be extended and applied far beyond the context of their original formulation. For example, the notion of a "gene" was first used by Hugo De Vries to characterize whatever "unit of inheritance" was implied by something similar to Mendel's theory, even though De Vries had not read Mendel and Mendel himself never used the term. Our understanding of the gene has since been greatly enriched and transformed by the advance of molecular biology, the mapping of the human genome, and so on. Contrary to the reductionist, there seems to be much more to genes than meets the eye.

The difficulty appears in behaviorism as well. When we say a person "believes it's raining" that seems to imply a potentially infinite set of possible behaviors – she brings her umbrella, wears a hat, curses the weatherman, etc. But are these behaviors really implied? If she also desires to get wet, then we should expect the opposite behavior. We might attempt to provide a behavioral analysis of "desires to get wet" but the same problems will arise. Furthermore many scientific terms seem to be explicable only in terms of *possible* rather than actual phenomena. Consider a term like "water-soluble." This can't apply only to objects actually dissolved in water since that would imply that the half packet of sugar that never makes it into my coffee is not soluble. Perhaps we could associate its solubility with the chemical properties in virtue of which it would dissolve *if* placed in water. But this seems to be reducing an observable notion to a theoretical one, not vice versa.

A final deep problem with reductionism is the assumption that we can clearly separate the empirical and theoretical terms of a scientific language, so as to reduce the one to the other. As we have already discussed, Kuhn and others maintained that observation in science is itself "theory-laden." For example, an astronomer may point to a photo taken from the Hubble telescope and say "look at this white dwarf at the edge of Andromeda." Such an observation presupposes a good deal of astronomical theory. It could be "cleansed" of theoretical content, of course, by referring instead to "this lighter patch at the edge of these dark lines." But it is hard to see how astronomy could be reduced to a collection of such clumsy statements and still remain a useful science.

Few contemporary anti-realists are wedded to instrumentalism or semantic reductionism. But a less radical empiricist alternative to realism has won considerable support in recent years. *Constructive empiricism* (CE), whose main champion is the Princeton philosopher Bas Van Fraassen, rejects the instrumentalist view that theories are neither true nor false, as well as the radical empiricist notion that theoretical terms are shorthand descriptions of observations. But CE retains the empiricist view that science does not require belief in those parts of theories that go beyond the observable. The aim of science according to CE is "empirical adequacy," i.e. truth about the observable rather than truth about the unobservable. For example, it might turn out that the standard model in particle physics perfectly accounts for the accelerator data and astrophysical phenomena to which they are applied. According to the constructive empiricist, we are then entitled to *accept* the standard model, use it in future research, fund it, etc.; but we are not required to *believe* that there really are quarks, leptons, and so on.

In response to CE, scientific realists have been particularly critical of the crucial distinction between observable and unobservable. First, the distinction is not a sharp one – it seems to be an arbitrary matter where we draw it. Presumably cancer cells and neurons, for example, do not count as observable even though they can be seen with the aid of a high-resolution microscope. But there seems to be a continuous scale from these to less powerful microscopes, then to magnifying glasses, ordinary eyeglasses, and the polarized windshield of my car. Surely what I see through my windshield is observable – so where to draw the line? Anti-realists answer that the absence of a sharp line does not invalidate the observable/unobservable distinction. The legal age to drive may be somewhat arbitrary but it does not follow from this that any toddler should be given a license. Butterflies and planets are clearly observable; electrons and dark matter are clearly not.

But even such clear-cut examples, the realist objects, depend on idiosyncratic facts about the sense organs of human beings. If our eyes were much sharper or implanted with microscopic technology (which may eventually happen as we'll see in chapter six), then the small-but-nearby could be observed. And conversely if we were rooted to the ground like trees then the large-but-distant could not be observed. In other words, the observable-unobservable distinction is relative. This is not really a problem for the constructive empiricist, however, since they are concerned with what *we* – members of our "epistemic community" – are entitled to believe, not science-fictional variations of ourselves. So what constructive empiricism amounts to is an empiricist form of skepticism about the unobservable, which will fall or stand with the standard realist and anti-realist arguments discussed above.

The forms of anti-realism we have considered so far share with realism the conviction that science aims to provide true and objective knowledge of the natural world. The anti-realist simply restricts our knowledge to the observable world. A final version of anti-realism, which we may label *conceptual relativism*, jettisons altogether the notion that science describes a world independent of our theories. We have already had a glimpse of such a view in Kuhn's radical suggestion that the "world changes" in a scientific revolution. In Kuhn's view, because the scientist's concepts and observations are both structured by the paradigm in which she works, there is therefore no "paradigm-neutral" way of characterizing the world of scientific inquiry. Furthermore, the conceptual divergence between paradigms is typically so drastic that there is no way to translate the terms and laws of the one theory into those of the other. The concept of time, for example, is invariant and independent of motion in Newton's physics, while in Einstein's physics time is variable and dependent on relative motion. The equally fundamental notions of mass and space are also very different in the two theories. Owing to this "incommensurability" among the concepts, observations, and methods of successive paradigms, Kuhn concludes that we must relinquish the traditional notion that there is one "full, objective, true account of nature" and that "changes of paradigm carry scientists and those who learn from them closer and closer to the truth."

Kuhn's relativism applies strictly only within science. And indeed Kuhn repudiated the more radical forms of relativism that his work inspired (see chapter five). But it is worth considering a more general form of conceptual relativism since, as we will see in the next chapter, relativist arguments have led many to conclude that all "ways of knowing" – scientific, religious, artistic – are equally valid. One of the few philosophers Kuhn mentions in *Structure of Scientific Revolutions* is Nelson Goodman. Applying the tools of logical analysis, Goodman tried to demonstrate that there are no absolute facts about the world independent of the "versions" or "descriptions" we make of it. For example, in his book *Ways of Worldmaking* Goodman argues that we can say whether the sun moves or not according to the geocentric vs. heliocentric theories of the universe, but we cannot ask simply whether the sun really moves or not.

> If I ask you about the world you can offer to tell me how it is under one or more frames of reference; but if I insist you tell me

how it is apart from all frames what can you say? We are
confined to ways of describing whatever is described. Our
universe, so to speak, consists of these ways rather than of a
world or of worlds.

A realist may concede that motion is dependent on description,
since whether a thing is moving or not might be an inherently
relational question (like whether a thing is large or small). For
example, even in the heliocentric model the sun moves relative
to other stars. But it is surely not a matter of description whether
the sun and moon are spherical or not, whether they are solid or
gaseous, lifeless or inhabited, and so on. Yet Goodman maintains
that these various ways of categorizing are themselves *made*
rather than simply discovered: "The many stuffs – matter,
energy, waves, phenomena – that worlds are made of are made
along with the worlds."

The radical thesis that the world is "made" by human
descriptions has dramatic implications that certainly merit
caution. For instance, if the stuff of which the sun and moon
consist is made by our descriptions, then it follows that the sun
and moon did not exist before humans described them.
Furthermore, if facts are relative to descriptions, then it seems
that it could be a fact both that the world is composed of one-
dimensional vibrating strings (relative to modern particle
physics), that it is composed of qi (according to ancient Chinese
medicine), and that it is composed of water (according to
Thales). Science would then offer no more valid a picture of the
world than mythology or fairy tales. Now, Goodman denies that
his pluralist view undermines science: "The pluralist, far from
being anti-scientific, accepts the sciences at full value." Kuhn
would agree. Nevertheless, the conceptual relativisms of Kuhn
and Goodman have encouraged the recently popular view that
science is a mere "social construction" with no special claim on
objective truth. Such views are a main topic of the next chapter.

Reduction and unification

Besides understanding the world itself, or at least the world as it appears to us, what other fundamental aims might science have? From the earliest Milesian cosmology, when Thales pronounced that everything is water, to the recent efforts to meld quantum theory and relativity, science has been enchanted by the dream of an ultimate "theory of everything." There are two dimensions to this dream, which we should treat separately: unification and reduction.

Unification occurs *within* a given science when two or more kinds of phenomena that had previously been covered by distinct concepts or laws are brought under a single analysis. For example, one of the monumental achievements of recent biology is the "modern synthesis" of Darwinian evolutionary theory and Mendelian genetics. Although, on the one hand, Darwin captured the mechanism of natural selection, he failed to identify either the means by which adaptive traits are inherited or the source of the random variation that powered evolution. On the other hand, following upon the work of Mendel, biologists had achieved a pretty clear understanding of genetic inheritance and variation in particular cases but had no way of relating this understanding to large-scale and long-term evolutionary change like speciation. In the 1930's the field of population genetics developed mathematically sophisticated accounts of genetic change that, in effect, showed how small-scale genetic processes combine to produce the large-scale evolutionary change familiar to field naturalists. Another important unification was achieved in the mechanics of the seventeenth century. Following Aristotle, physics had been guided for hundreds of years by the assumptions that different natural laws obtain in the celestial vs. the terrestrial (sublunary) realms. In particular, it was assumed that celestial motion is naturally circular and eternal while terrestrial motion is naturally toward the center of the earth and final rest. The detailed investigations of Galileo and Newton eventually led to the realization that all motion is in itself constant and rectilinear, and only otherwise owing to the presence of specific forces. The motions of planets, projectiles, and pendulums became local manifestations of the universal essence of motion.

Reduction is the attempt to show that the concepts and laws of a given science follow directly from the concepts and laws of another, more fundamental science. Typically, the "more fundamental" science deals with smaller and more universal entities and processes. For example, a reductionist may try to show that psychology is reducible to neurobiology since all brains (as far as we know) are composed of neurons, or that chemistry is reducible to physics since chemical processes and structures are composed of physical processes and structures like atoms. Another kind of reduction occurs not between sciences but among different levels within the same science. For example, Newton was able to show how various specific laws that had already been accepted within mechanics – Kepler's law of the orbits, Galileo's law of falling bodies, etc. – followed from the more general Newtonian laws of motion and universal gravitation.

As with unification, reduction in particular cases can bring considerable light to an area of inquiry and suggest avenues for future research. For example, beginning in the seventeenth century the basic relationship among the temperature (T), volume (V) and pressure (P) in gases was charted experimentally and discovered to conform to the following law:

$PV = rT$

where r is a constant.

In the nineteenth century it was shown that this "ideal gas" law followed from the kinetic theory, which conceived of gases as tiny corpuscles colliding at various rates in accordance with the laws of thermodynamics and Newtonian mechanics. The law describes an "ideal" gas because certain idealizations are required to achieve the derivation. For example, the corpuscles must be regarded as perfectly elastic and point-sized. Nevertheless, this reduction in effect explained *why* the gas law obtains. So successful unification and reduction have usually been welcomed in science for bringing coherence and perspective. Failure can be instructive, too. For example, Descartes attempted to show that gravity followed from his fundamental view that all of space was full of bodies of different sizes interacting by direct contact. But neither Descartes nor his followers were able to produce a plausible reduction and this failure hastened acceptance of Newton's non-reductionist theory of gravity.

Beyond their use in guiding particular research programs, unification and reduction have sometimes been promoted as broad ideals of the final aim or meaning of science. Many of the most prominent philosophers of science in the early twentieth century were committed to a "Unity of Science" movement, the hope of which was, as the editors of the *Encyclopedia of Unified Science* put it, to "reduce all scientific terms to one kind of term using a special logical technique." But wholesale reductionism has faced numerous philosophical difficulties. Although based on the plausible assumption that the complex systems investigated in sociology, psychology, and biology are ultimately composed of physical entities and processes, it is not clear that the *laws* and *concepts* found in the so-called "special sciences" must be reducible to those found in the physical sciences. For one thing, the more complex processes are often context-sensitive. How a gene expresses itself in a given organism, or how a mental illness develops in a given person, will depend on numerous factors that are not simply physical facts about genes or neurons, including the presence of other biological factors in the organism and the person's social environment.

Furthermore, it seems likely that the laws and processes of the special sciences may be "multiply-realizable," meaning that very different kinds of physical systems will equally satisfy those same laws. The philosopher Jerry Fodor gives the economic example of Gresham's law, which states that when there are two forms of currency in circulation (e.g. paper and gold) the one with less commodity value will eventually dominate (i.e. paper). Fodor points out that such a law, assuming it holds, will be satisfied by countless forms of currencies, including coins, paper, checks, grain, and "wampum." The physical facts about these currencies are very different and yet they all satisfy Gresham's law, suggesting that the law is not ultimately reducible to physics. In psychology, multiple-realizability is crucial to the question of whether organisms with brains very different from ours might nevertheless have similar minds.

It is fair to conclude that the value of unification and reduction should be decided on a case-by-case basis, by nature itself rather than by a priori philosophical ideology. It may even turn out that some natural systems have features that are genuinely *emergent* from the things that ultimately compose them. Consciousness, for example, may be emergent from the brain in the sense that it has features like intentionality and self-awareness that cannot be found at the level of neurons or molecules. Emergentism in this sense need not imply that consciousness is distinct from the brain, only that there are properties caused by the brain that are not themselves strictly neurochemical. Some emergentists hold, in addition, that the higher levels of systems have causal powers of their own which can exert influence "downward," as when inflationary trends in an overall economy cause anxiety at the level of an individual consumer which in turn causes some nerve and muscle actions when the person decides to spend cash before prices rise.

Another alternative to reductionism is *pluralism*: the acceptance and encouragement of a multiplicity of independent theoretical perspectives on a given natural system. Mental illness, for example, can be studied from both a cognitive and a neurophysiological point of view and can therefore be treated without having to choose between therapy and medication. But in other contexts pluralism may be more difficult to tolerate. In fundamental physics for example, the search for a theory of everything is not motivated only by an obsession with unity, but rather by a serious inconsistency between general relativity and quantum theory. Each theory, as things stand, characterizes the physical world in a way the other precludes. Still, as we will see in the next chapter, pluralism can be a useful stance when examining the social and political dimensions of science. For despite recent acrimonious controversy about postmodern "deconstructions" of science, it may turn out that science is both a social construction *and* our truest, most objective picture of the world.

5

The social dimensions
of science

Science has always been a social phenomenon. It's a human activity after all, and humans are, as Aristotle observed, "by nature a social animal." Even the mythical solitary genius, or "mad scientist," must rely on others for an education. Two of the greatest figures of the Scientific Revolution, Descartes and Newton, prized their solitude (for this reason Descartes changed addresses frequently) but drew heavily on the work of others through correspondence and reading. Neither one was especially gracious about acknowledging their intellectual debts, although Newton famously conceded in a letter to his fellow scientist Hooke, with whom he later quarreled bitterly, "If I have seen further, it is by standing on the shoulder of giants." Nor were these early scientists isolated from the tumultuous social changes that surrounded them: the Thirty Years War and the Inquisition in the case of Descartes and the English Civil War and Glorious Revolution in the case of Newton. The unsettled intellectual and cultural milieu of seventeenth-century Europe – the schisms within Christianity, the rise of global trade, national and religious wars – profoundly influenced the early development of modern science.

Science is part of society and therefore subject to social forces. Increasingly, science itself is a social process. Copernicus, Newton, and Descartes worked in semi-isolation, as did Einstein in the early years of his career. But today a typical laboratory at a university or private research institution may employ dozens or

even hundreds of scientists with specific duties. Arrays of technical assistants are also required to support the complex technologies typically involved in advanced research. Amateur scientists
and independent scholars played an important role in the growth
of modern science (Joseph Priestley and Gregor Mendel are two
good examples); but today almost all scientists have Ph.D's and
academic or industry affiliation. Funding is provided by public
granting agencies, like the NSF or NIH in the United States, as
well as private institutes and corporations, each with their own
agendas and responsibilities. Research is evaluated for publication though a deliberative system of peer review and then
disseminated in highly specialized professional journals. In many
fields a research article will have more than twenty authors,
arranged hierarchically like film credits. Like the movies
produced by a Hollywood studio, or the laws enacted by a
legislative body, scientific knowledge is the product of a
complex social web.

Social constructivism

So it is not surprising that the social dimensions of science have
been closely studied, especially in the last century as the social
sciences themselves have matured. Originally, the focus of study
in the sociology of science was the social impacts of the technological advances brought by science. But sociologists soon
turned to the social aspects of science itself. One of the pioneers
in this field, Robert K. Merton, developed the theory now
known as *Merton's thesis* that the social norms associated with
German and English Protestantism – communalism, universalism, disinterestedness, originality, and skepticism (CUDOS) –
were a major impetus for the emergence of modern science in
Western Europe. In this view, science was inseparable from its
social context.

Competing sociological schools proposed alternative social explanations for science. Marxists, for example, emphasized economic forces while other schools explored the link between science and the rise of democracy. Generally, however, these early sociological approaches were more concerned to explain in social terms the origin of science and its institutional structure rather than its methods and content. In terms of the movie industry analogy, one could say they attempted to explain the rise of the Hollywood studio system, and its means of funding films, rather than the styles and subjects of the films produced. Furthermore, the early sociologists tended to assume science possessed a structure and dynamic that was distinct from society as a whole, which was the source of its unique achievements.

More recent sociology of science has been strongly influenced by the work of Thomas Kuhn. Although Kuhn stressed certain parallels between political and scientific revolutions, and the somewhat dogmatic and paradigm-centric nature of scientific education, he did not discuss in detail the social forces affecting science. Rather, like Merton, he stressed the unique power of science, especially "normal science." For Kuhn, normal science is a social phenomenon but its "insulation" from the broader society was essential to its technical efficiency and progress. However, sociologists were encouraged by Kuhn's assault on the prevailing logical models of scientific method. If history showed that purely rational considerations played only a minor role in major scientific changes, then the door was open to alternative, especially social, explanations. The new Sociology of Scientific Knowledge (SSK) sought to explain the results of science, its theories and technologies, as socially determined no less than its dynamics and organizational structure.

Some early forms of SSK were restricted to the explanation of scientific errors. For example, the persistence of classical ideas in some parts of the physics community in Nazi-era Germany, despite the clear superiority of quantum theory, was attributed to misplaced antipathy to the supposedly "Jewish science" of Einstein and Bohr. The presumption was that while errors might have social explanation, the cause of successful science could only be the rational pursuit of truth. However, in the 1970s a group of scholars at Edinburgh and Bath proposed to replace this "weak" programme in SSK with a "strong programme." David Bloor offered a kind of manifesto for the strong programme, a central plank of which was the following thesis:

> **Symmetry Thesis:** SSK should be "symmetrical in its style of explanation. The same types of causes would explain true and false beliefs."

Just as in the sociology or anthropology of religion, presumptions about the truth or rationality of scientific practices or beliefs should play no part in their causal explanation.

The strong sociological approach to scientific knowledge has been applied to many historical case studies, uncovering the role of social forces and interests (economic, political, colonial, gender) in important scientific controversies and discoveries. A notable recent example is the study by Steven Shapin and Simon Schaffer of the experimental investigation of the vacuum in seventeenth century England, *Leviathan and the Air Pump: Hobbes, Boyle and the Experimental Life*. The debate between the philosopher Thomas Hobbes and the scientist Robert Boyle ostensibly concerned a straightforward scientific matter – the existence of vacuums. Boyle thought they could be produced in the laboratory while Hobbes defended the traditional view that "nature abhors a vacuum." Shapin and Schaffer show that the controversy was inextricably tied up with broader questions about the proper methodology of the newly emerging science. Boyle advocated a thoroughly experimental approach while Hobbes preferred logical and theoretical reasoning. Accordingly, Hobbes dismissed Boyle's experiments as uncertain and artificial distortions of nature, irrelevant to the fundamental question concerning vacuums. This methodological dispute was itself a manifestation of pressing political uncertainty concerning the control of knowledge in the wake of the English Civil War. The experimental approach, associated with Boyle's Royal Society of London, stressed a collaborative and fallibilist conception of scientific knowledge. This reflected, on the one hand, their support for democratic, parliamentary rule over absolute monarchy and, on the other hand, their opposition to the unyielding dogmatism of religious enthusiasts. But Hobbes, who had defended absolute monarchy in *Leviathan*, his famous work in political philosophy, feared that "rule by consensus" was a recipe for error in science, and a return to chaos and strife in society. He also suspected that the experimentalists' privileging of consensus among trained experts in science was a ruse for the suppression of intellectual dissent.

Though purely factual on its surface, the causes, content and final resolution of the Boyle-Hobbes debate cannot be understood independently of the role each side played in the turbulent social and political events of Restoration England. From this, Shapin and Schaffer draw a quite provocative conclusion about scientific knowledge generally. The facts about the vacuum, about the natural world, are not different in kind from the facts about our customs, our laws, and our political loyalties. Natural facts are not simply "out there" waiting to be discovered, but are instead constructed or manufactured in a complex social nexus: "it is ourselves and not reality that is responsible for what we know." So let us next examine this notion that scientific knowledge is ultimately a social construction.

Strictly speaking, the Strong Programme should be neutral on the traditional philosophical issues of realism and relativism since it is concerned with the causes of scientific beliefs, not their truth or falsity. Nevertheless, the Strong Programme's proponents have generally been relativistic in orientation and overtly hostile to the realist and rationalist stance of mainstream philosophy of science. Besides Kuhn, many advocates for the Strong Programme drew on French philosophers like Foucault and Bachelard, who were more interested in the historical construction of scientific authority in relation to broader social and psychological forces than in abstract questions about the essence of method and progress. They were also impressed by the philosopher Wittgenstein's critique of philosophy's excessively universalistic and static approach to meaning. According to Wittgenstein's alternative picture, meaning is determined by use in linguistic communities – what he called "language games" – not by correspondence to Platonic forms or essences. These various tendencies culminate in the view known as "postmodernism," which rejects the traditional conception of knowledge as a subjective representation of objective reality, or a "mirror of nature." Associated values of modern philosophy, such as realism, objectivity, rationalism, and commitment to progress are also jettisoned. Thus, postmodernists like Jean-Francois Lyotard reject all "grand philosophical narratives" about science, such as realism or inductivism, and focus instead on historically and culturally situated practices.

An interesting approach that combines the sociological method with the postmodernist abandonment of traditional philosophy is the 1979 book *Laboratory Life: The Social Construction of Scientific Facts*, by the British sociologist Stephen Woolgar and the French philosopher Bruno Latour. Woolgar and Latour studied a modern cell biology lab (the Salk Institute in California) in the way anthropologists might study an indigenous culture, closely observing and cataloging the practices and interactions among the researchers from the inception of a project to the publication of final results. They even constructed maps of the laboratory to track the movements of its "natives" over the course of a day. Far from the logical abstraction of Popper and Hempel, and more detailed and specific than Kuhn's description of normal science, Latour and Woolgar depict science as a complex network of competition, negotiation, material exchanges, data massaging, and status-mongering. Like Shapin and Schaffer, they assert that scientific facts (in this case about the human hormone TRF) are not discovered or represented but rather constructed: "It was not simply that TRF was conditioned by social forces; rather it was constructed by and constituted through micro-social phenomena."

The social constructivist analysis of science spurred a more encompassing, interdisciplinary approach known as "science studies," or "science and technology studies," which brought together historical, anthropological, literary, post-colonial, and feminist perspectives on scientific practice. Science studies grew in popularity in the 1970s and 80s and became somewhat associated with "cultural studies," an even broader, decidedly postmodern, disciplinary orientation with roots in critical theory and leftist politics. Unsurprisingly, science studies found little support from traditional philosophy, although many philosophers of science were sympathetic with the critique of extreme rationalism. Mainstream scientists generally dismissed the social constructivist view of science as ill-informed and obscurantist. There was also suspicion that science studies faculties were anti-science and determined to undermine the considerable prestige and authority of science. This roiling academic controversy came to a head in 1996 when Alan Sokal, a physicist from New York University, published an article in a science-themed issue of *Social Text*, a major cultural studies journal. Sokal's article was called "Transgressing the Boundaries: Towards a Hermeneutics of Quantum Gravity." A skeptical observer of cultural studies, Sokal mimicked the style and jargon of the discipline in his rambling and nonsensical, but linguistically forbidding, pseudo-analysis of cutting-edge physics. Here is a brief sample from the article's introduction:

> In quantum gravity, as we shall see, the space-time manifold
> ceases to exist as an objective physical reality; geometry becomes
> relational and contextual; and the foundational conceptual
> categories of prior science − among them, existence itself −
> become problematized and relativized. This conceptual revolu-
> tion, I will argue, has profound implications for the content of
> a future postmodern and liberatory science.

Simultaneously with the *Social Text* article, Sokal published a
self-exposé in the academic magazine *Lingua Franca*. He
explained that his aim was to reveal the intellectual bankruptcy
of the cultural studies critique of science − "evidently the editors
of *Social Text* felt comfortable publishing an article on quantum
physics without bothering to consult anyone knowledgeable in
the subject" − but also, and more importantly, to underscore the
intellectual and political perils of constructivist thinking:

> While my method was satirical, my motivation is utterly
> serious. What concerns me is the proliferation, not just of
> nonsense and sloppy thinking per se, but of a particular kind of
> nonsense and sloppy thinking: one that denies the existence of
> objective realities, or (when challenged) admits their existence
> but downplays their practical relevance. At its best, a journal
> like *Social Text* raises important questions that no scientist
> should ignore − questions, for example, about how corporate
> and government funding influence scientific work.
> Unfortunately, epistemic relativism does little to further the
> discussion of these matters.

The hoax generated considerable media attention worldwide −
scientific realism is hardly a common topic in the international
press − and triggered an explosion of debate that was dubbed the
"science wars." The editors of *Social Text*, after apologizing to
their readership, explained that the journal has always embraced
an "editorial milieu with criteria and aims quite remote from

those of a professional scientific journal" – although the issue contained many eminent science studies scholars, as the editors go on to note. They also disparaged the "deceptive means" by which Sokal chose to make his point, as did Stanley Fish, the prominent literary theorist and then director of the publisher for *Social Text*, Duke University Press. Veterans of the science wars like the physicist Steven Weinberg took the hoax to reveal "a problem not just in the editing practices at *Social Text*, but in the standards of a larger intellectual community." The intellectual community he's speaking of is science studies.

The response of philosophers of science was more ambivalent. On the one hand, many welcomed Sokal's lampoon of the carelessly relativistic, sometimes scientifically illiterate, work of certain science studies writers. The realism debate has a long and distinguished history in philosophy, but the postmodernists' position was based only on obscure metaphysics, fashionable politics, and a misreading of Kuhn. (Indeed Kuhn himself wrote in a retrospective essay called "The Road Since Structure," "I am among those who have found the claims of the strong program absurd: an example of deconstruction gone mad.") On the other hand, philosophers worried that Sokal's over-the-top ridicule might undermine the genuine insights that have been offered by the social approach. Thus, in his "Plea for Science Studies," Philip Kitcher called attention to the "significant number of important articles and books" by recent science studies scholars offering a perspective on the practice of science that the practitioners themselves miss.

So what lessons should be drawn from the Sokal affair? Weinberg seems right that the issue is not so much the integrity (or lack thereof) of the peer review process employed by cultural studies journals. The editors of *Social Text* might take some comfort in the fact that even physics journals are susceptible to frauds and hoaxes. Recently two brothers, Igor and Grichka Bogdonov, published articles in respected particle physics journals that have since been widely judged by experts as incoherent jumbles of technical buzz-words. In 2005, two MIT computer science graduate students had a gibberish paper generated by a computer program – "Rooter: A Methodology for the Typical Unification of Access Points and Redundancy" – accepted for an international conference. Thousands of academic journals and conferences review countless articles every year; it would be astounding if imposters did not occasionally slip past the firewall of peer review.

But Sokal, like Weinberg, takes the acceptance of his paper as evidence of the failure of cultural studies as a whole. In his view the vetting process at *Social Text* failed precisely because cultural studies adopts a relativistic and postmodern attitude toward traditional notions of truth and objectivity. The ease with which he was able to pass off his poppycock as serious scholarship, apparently without any external review by a scientist, reveals to what extent postmodern critiques of science have abandoned intellectual rigor and standards. Sokal writes, "No wonder they didn't bother to consult a physicist. If all is discourse and 'text', then knowledge of the real world is superfluous; even physics becomes just another branch of Cultural Studies. If, moreover, all is rhetoric and 'language games', then internal logical consistency is superfluous too: a patina of theoretical sophistication serves equally well."

The editors of *Social Text* could plausibly respond to this charge in two ways. First, it does not follow from the fact that the article was sloppily reviewed that this was *because* the governing epistemology of the journal is relativist. The editors may have been sloppy for the same reasons the editors of the physics journals and computer science conferences were. As Sokal concedes, since his experiment is "uncontrolled" – i.e. there is no way of knowing whether the paper would have been accepted by non-relativist editors in similar circumstances – the question of the real cause is underdetermined. Second, and more importantly, the editors could simply dismiss Sokal's appeal to traditional standards of intellectual rigor. If all really is social construction and language games, they could just refuse to play Sokal's "language game" of intellectual rigor and peer review. But as we shall see in a moment, though this sort of response is logically consistent, it brings considerable difficulty for the social constructivist.

Although Sokal's hoax does not in itself invalidate science studies, it certainly does raise serious worries about the tenability and coherence of the most radical forms of social constructivism. First, constructivism has strange implications for our common-sense view of the world over time. It seems that in whatever sense oxygen and TRF are constructed in the laboratories of modern chemistry, phlogiston and caloric were in the same sense constructed in the laboratories of nineteenth-century chemistry. So the world has changed: a candle burning used to emit phlogiston but now it consumes oxygen. And TRF was not in our brains before it was constructed in the laboratory by the Salk Institute researchers. And in the thirteenth century the sun and planets really did circle the earth on crystalline spheres since that was how natural philosophers constructed the heavens. Such views are most bizarre when applied to sciences that are concerned with the remote past. To say the Pre-Cambrian Era or the big bang are socially constructed makes no sense, since part of what scientists *mean* by these concepts is that they existed long before there was any science. Indeed it seems almost to be part of the "language game" of most science that its objects are not socially constructed. But then, in the constructivist view does that mean they're constructed or not? Whatever the answer, it seems social constructivism requires taking much too literally Kuhn's metaphorical remark about the world changing when science changes.

Most versions of social constructivism are not so radical as to say that electrons exist if and only if scientists believe they do.

Even Latour, reflecting recently on the "fact" of global warming, has seemed to concede an inherent "solidity" to some scientific facts: "critique was useless against objects of some solidity. You can try the miserable projective game on UFO's or exotic divinities, but don't try it on neurotransmitters, on gravitation, on Monte Carlo calculations." A more moderate form of constructivism might hold not that gravity and electrons depend for their *existence* on scientific theories but that their particular *natures* are a consequence of the scientific process through which they are investigated. Obviously, electrons and gravity existed in ancient times but they did not exist for Aristotle *as* they do for the physicist of today. Plain old rocks really are out there to be studied by scientists (and thrown by delinquents) but, as Ian Hacking has suggested, *dolomite* depends for its existence on the social and theoretical dynamics of modern geology. The earlier, more radical version of Latour once said that the pharaoh Ramses II could not have died from tuberculosis (as some have speculated from the condition of his mummified corpse) any more than he could have died from a machine-gun. What Latour might now say is that Ramses died of an infection with some "solidity" that modern medical science has since endowed with a complex matrix of theoretical properties.

But this more moderate form of constructivism still faces a fundamental problem, which relates to another plank of the strong programme in the sociology of scientific knowledge:

> **Reflexivity Thesis:** "In principle its pattern of explanation would have to be applicable to sociology itself."

This raises a difficulty with which philosophers are fond of tarring relativistic doctrines. Suppose someone baldly asserts "everything is relative." We may then ask whether "everything" includes the assertion just made. If the answer is "no," we are owed an explanation why this assertion alone gets to be absolute. If the answer is "yes," then the assertion is no real threat to traditional realism. For then it merely expresses the perspective of the relativist, a "social construction" or "language game" the realist can ignore as an idiosyncratic but harmless muddle. Either way, the bite is taken out of relativism. A similar problem afflicts social constructivism. Does this doctrine depict science as it really is, or is the social construction of scientific facts itself a mere construction? If the former, it's not clear why sociology alone among the sciences is capable of getting at the real world. But if social construction is itself a social construction, then realists are free to dismiss it as a mere fancy with as little relevance to real science as the science fiction produced in the creative writing department.

I think the best way for the social constructivist to respond to this problem is to say that their doctrine is *both* a social construction *and* true. It emerged from the complex social and political matrix of late twentieth-century academia influenced by the paradigm of sociology or science studies. For all that, it may reflect a correct perception of science. At least, contained in this perception is a genuine aspect of scientific practice that is not evident to the practicing scientists themselves, as Kitcher urged in defense of science studies. This is a sensible solution to the reflexivity problem: social constructivism is a social construction that happens to be partly true.

But this solution comes at a price. For it removes much of the threat to realism, and traditional philosophy of science, that social constructivism was alleged to pose. If theories and facts in science studies can be both constructed and true, then why can't theories and facts in biology and physics? The realist may decide to admit that scientific knowledge is largely a product of complex social forces and yet for all that insist that it captures the way things really are. Indeed, a moment of reflection will reveal that for most of us "socially determined" and "true" or "rational" coexist quite happily. Our religious beliefs (or doubts) may be largely a function of contingent social circumstance. Had we been raised by different parents in a different culture our religious views might have been very different. But this does not seem to prevent us from believing deeply that our religious beliefs (or doubts) are true and reasonable.

Even if things must be either real or constructed (either black or white), it doesn't follow that they must be entirely one or the other (black or white all over). It might turn out that we should regard some scientific facts as mostly constructed and others as mostly independent of us. Consider a psychiatric disorder like attention deficit/hyperactivity disorder (ADHD) which denotes a constellation of behaviors like restlessness and inability to concentrate. Is ADHD a social construction? One might be inclined to say "no" considering that many patients receive measurable benefits, like improved academic performance, from the drug Ritalin. Then again, so do people without symptoms of ADHD.

One useful way of approaching the question of social construction, suggested by Ian Hacking, is to ask whether it was inevitable that scientific inquiry should have arrived at the theory under consideration even in very different social contexts. We regard current fashions and sports as social constructions because we can easily imagine reasons why we might have had different interests in these things. In the case of ADHD it is arguable that psychiatry would not have conceived this disorder if we had no public education system and children simply played freely until adulthood. It is harder to conceive counterfactual social circumstances that would have resulted in chemistry without hydrogen and oxygen. These concepts are so central to modern chemistry that if for some reason they became socially disfavored we would have not a different chemistry but no chemistry at all. At any rate, there seems to be no reason to decide in advance whether a fact is constructed or independent. One of the important lessons of recent social studies of science is that the question of social construction is a matter for careful historical and empirical investigation, and not to be legislated by philosophical fiat.

Feminist philosophy of science

With a few notable exceptions, since the seventeenth century women have played only a marginal role in science. That situation changed quite dramatically in the second half of the twentieth century. Women now represent large minorities in fields such as psychology, biology, and the social sciences, though they still hold less than twenty percent of university faculty positions in physics and astronomy. The reasons for women's exclusion from science are partly the same as for their historical exclusion from other powerful, high-status professions, such as law and the clergy, which are reserved for the ruling groups in oppressive social systems. In addition to these obvious political causes, feminist scholars have also suggested that the longstanding dualistic opposition in philosophy between, on the one side, women, the body, and emotion, and, on the other side, men, the mind, and reason, has encouraged the assumption that women are not suited for careers in science and philosophy. It will be useful to keep this in mind as we ask what impact the exclusion of women has had upon science, and how science might be changed as women become more involved.

On some conceptions of science, the answer to these questions should be "none at all." If science is a purely rational process of data gathering and theory testing, then the gender of its practitioners should have no more relevance in science than in accounting. Whether a theory is empirically successful, it might seem, has no more to do with gender than whether a spreadsheet balances. While there may be good moral and practical reasons for promoting greater participation by women in science, given the prestige of scientific careers and the intellectual resources that are otherwise squandered, these reasons are "external" to the practice and content of science itself.

However, with the decline of the logical empiricist model of science scholars began to examine more closely the ways in which factors like gender (along with race, class, and ethnicity) influence science. There seem to be three possible kinds of influence. First, and most obviously, gender might affect the *selection and prioritization* of research problems as well as the practical application of scientific knowledge. For example, had there been more women in medical science, research on the risks associated with pregnancy, such as pre-eclampsia and postpartum depression, might have been investigated more thoroughly. And, arguably, basic research in fields like physics and biology might have been directed into applications in agriculture and medicine more than the military and industry. Second, gender might influence the *practice and methods* of science. It may be, for example, that women tend to prefer, or are generally more experienced and better trained in, collaborative rather than individualistic, competitive approaches to problem solving. Or it may be that women generally prefer analogical and model-based analyses of empirical phenomena while men tend to use more quantitative and formal representations.

Finally, and most controversially, gender may enter inextricably into the *content* of scientific knowledge in such a way that our theories of the world might have been very different if women had played an equal or greater role than men. For example, less atomistic and more holistic models of the structure of matter might have been more prevalent had women played a greater role in the early stages of physics. And since there is some evidence that women are more tolerant than men of diverse and partial representations of a given phenomenon, there may have been less emphasis on reductionism in biology and psychology.

I have said "may have" frequently in the above summary of the influence of gender in science. It is difficult, and potentially misleading, to generalize about this influence without taking careful account of the historical (as well as sociological and psychological) evidence we possess. Some of the more influential analyses of specific ways in which gender has influenced science include: Carolyn Merchant on the history of misogyny in science in *Death of Nature,* Helen Longino on the role of gender in theories of brain development in *Science as Social Knowledge*, Ruth Bleier on sexism and biology in *Science and Gender: A Critique of Biology*, Donna Haraway on gender in the history of primatology in *Primate Visions*, Alison Wylie on archeology in *Thinking from Things*, Sarah Hrdy on evolution and anthropology in *The Woman that Never Evolved*.

Let's consider one case which has been examined closely by the feminist scientist and author Evelyn Fox Keller. Her subject is the eminent twentieth-century geneticist Barbara McClintock, whose groundbreaking discoveries in the complex internal dynamics of gene regulation earned her a Nobel Prize in 1983. Keller relates the research style that fostered McClintock's discoveries – patience, attention to detail, unorthodoxy – to McClintock's own characterization of her intuitive conceptions of nature – an empathetic "feeling for the organism," a sense of the oneness of things, a belief in the fathomless complexity of biological systems, and so on. Contrasting the dominant "hierarchical" view of gene dynamics in which a "master-molecule" directs subordinate activities, with the "non-hierarchical" view in which "control resides in the complex interaction of the entire system," Fox Keller argues that McClintock's adoption of the latter, less "masculinist" orientation allowed her to comprehend the delicate mutual interactions involved in gene regulation. On the down side, her method also closed her off from further theoretical advances when the putative master molecule (DNA) was discovered by Watson and Crick (with the help of Rosalind Franklin's x-ray pictures).

A feminist might reasonably worry that Fox Keller's championing of McClintock's "intuitive" and "sympathetic" style of scientific reasoning will only reinforce sexist stereotypes that contrast the supposed emotional subjectivity of women with the rational objectivity of men. For this reason, Keller herself cautions against "rejecting objectivity as a 'masculine ideal' since that . . . exacerbates the very problem it wishes to solve." Instead, she says, "the first step in extending the feminist critique to the foundations of science is to re-conceptualize objectivity." What then would a feminist notion of objectivity entail, and what underlying feminist vision for science would it support?

Assuming that gender has entered into science in some or all of the ways mentioned, resulting in a view of nature that reflects the interests, values, and perspectives of men more than women, what is the alternative? How exactly might a "feminist science," or simply a non-sexist science, differ from male-dominated science of the past? One view is that feminist science is simply science cleansed of the male bias (in research prioritization, method, and content) that has distorted science. This conception, sometimes called "feminist empiricism," retains the traditional notion of objectivity as gender-neutral and disinterested, but discredits "masculinist" science for betraying this valid ideal. Consider, for example, feminist critiques of biological theories of human reproduction. Going back to Aristotle, the egg has been conceived as playing a mostly passive, "nurturing" role while the sperm is seen as highly competitive and active. Feminists have plausibly suggested that such models replicate the traditional sexual and social roles of men and women and consequently obscure understanding of the complex contribution of the egg and uterus to the early stages of reproduction. For similar reasons, biologists have been slow to appreciate the sometimes-competitive relationship between the fetus and the mother, and the frequency of parthenogenesis (fertilization without the male) in large animals. The lesson, according to feminist empiricism, is not to replace the male perspective with a female one – perhaps one which has the egg doing all the hard work while the sperm puts its feet up – but rather to move beyond such perspectives to a genuine, gender-neutral objectivity.

Although feminist empiricism seems quite attractive in principle, many feminist philosophers of science have remained dubious of any return to the ideal of a completely value-neutral empirical foundation for science. They take to heart the lesson of Kuhn and the social constructivists, as well as Quine, that there is no absolutely objective guide to theory choice. Furthermore, feminists with political or social aims would like to find a way for values and social perspective to enter into theory-choice, though in a way that does not merely distort research in the opposite direction from male-biased sexist science.

One of the most sophisticated versions of feminist empiricism that attempts to incorporate political and social values into science is developed by Helen Longino. Relying on a version of the empirical underdetermination argument discussed in chapter four, Longino argues that there is an unavoidable logical and evidential "gap" between our theories and the evidence we possess. Since the data itself cannot dictate a choice between empirically underdetermined theories, this choice must fall back upon "background assumptions" that involve implicit values. The most "objective" way of applying background assumptions in theory choice, according to Longino, is to rely upon the critical reflection of the community of scientific researchers, guided by shared standards and equality of intellectual authority, together perhaps with input from members of the broader community. This democratic approach to theory-choice seems preferable to relying on the background assumptions of a privileged few, since these may be arbitrary, subjective and self-serving.

A somewhat more radical conception of feminist science seeks to displace altogether the traditional "God's eye" view of objectivity on the grounds that this has merely served as a mask for the perspective and interests of dominant groups. In sympathy with the postmodernist rejection of the traditional "mirror of nature" model, some feminists maintain that there is no univocal "take" on reality, only the situated knowledge of subjects with particular histories, social locations, and cognitive styles. Others develop the originally Marxist idea that the "standpoint" of the socially disadvantaged gives them an epistemic advantage just as the "slave" understands social inequality better than the "master." What these approaches have in common is the conviction that science from the point of view of only the dominant social groups is at best partial and at worst delusional. Thus, Sandra Harding has proposed substituting the "weak" conception of objectivity associated with traditional epistemology with a "strong objectivity" that takes into account the situated perspectives of the knower herself and privileges the perspectives of the disadvantaged, such as women: "Starting research in women's lives leads to socially constructed claims that are less false – less partial and distorted – than are the (also socially constructed) claims that result if one starts in the lives of men in the dominant group."

But what remains within feminist science of the traditional aim of truth? Feminists share with anti-realists, especially social constructivists, deep misgivings about the dream of one true theory of everything. Knowledge is too laden with social values and cognitive biases to approximate the rationalist ideal of absolute (weak) objectivity. One might substitute for this a purely pragmatic account of truth as whatever best serves the social and material aims of a given epistemic community. But this is an inadequate characterization of the aspirations of feminist science. It does not merely serve the ends of feminism (as sexist science served the ends of men). Feminists are also committed to the realist hope of achieving better, "less distorted" knowledge of the world than dominant science produces by exposing its arbitrary background assumptions and by including and combining multiple perspectives on the natural world. It seems, therefore, that an aim of feminist science – or of any social movement that aims to make science better – is not absolute truth but rather increasing "truthlikeness," as characterized in chapter four. Furthermore, within this "progressivist" epistemology, the feminist can promote a pluralist, anti-reductionist toleration for multiple, complementary accounts of the same natural systems. As Longino puts it: "Nature may be so complex that it is impossible for any given account of a process to represent fully all the different processes that make a difference to the precise course of that process." While eschewing a single or final truth about nature, we can still make real and steady progress.

Science and values

Both social constructivists and feminists maintain that science is shot-through with values. This represents a major challenge to traditional philosophical thinking, since within philosophy the notion that facts (how things are) and values (how things ought to be) are completely separate issues has a status approaching dogma. The classic expression of this strict is/ought dichotomy is given by David Hume. If it *is* the case that the climate is changing, and it *is* the case that CO_2 is the main cause, does it follow that we *ought* to reduce carbon emissions from cars and factories? Not at all, according to Hume, since it is "altogether inconceivable how this new relation can be a deduction from others which are entirely different from it." Nor does it follow that we *ought not* to reduce carbon emissions from cars and factories. Hume's point is that reasoning about facts can't dictate moral preferences either way: "it is not contrary to reason to prefer the destruction of the whole world to the scratching of my finger." In one form, the derivation of moral conclusions from factual premises has been labeled the "naturalist fallacy" (which was discussed near the end of chapter three): holding that something is good (or bad) because it is natural (or unnatural). For example, vegetarians sometimes argue with meat-eaters about whether humans are "naturally" carnivorous. The vegetarians point out that we have a long digestive track that is ill-suited for the processing of meat. The meat-eaters reply that we have canine teeth which other mammals use to tear flesh. Both sides are stuck in the naturalist fallacy: nothing follows about the morality of meat-eating from its being natural (or not).

What about drawing inferences in the other direction, from values to facts? Philosophers have discussed this less frequently, perhaps because it seems even more obviously fallacious. We ought to drive at the speed limit and love our neighbor, but most of us do neither. The philosopher Immanuel Kant thought that God must exist since otherwise there is no guarantee that morality coincides with happiness in the grand scheme of things. About this kind of reasoning the philosopher Bertrand Russell remarked, "If you looked at the matter from a scientific point of view, you would say, 'After all, I only know this world. I do not know about the rest of the universe, but so far as one can argue from probabilities one would say that probably this world is a fair sample, and if there is injustice here then the odds are great that there is injustice elsewhere also.'" Russell doesn't do justice to the Kantian argument. But we can see his point: insomuch as science is concerned strictly with the facts, considerations of values like justice simply don't enter in.

The fact-value distinction was upheld in pre-Kuhnian philosophy of science, especially by the logical empiricists. Indeed the early logical empiricists thought that ethical (and metaphysical) claims are not really meaningful at all since they cannot be empirically verified. Rudolph Carnap, for example, maintained "All statements belonging to Metaphysics, (regulative) Ethics . . . are in fact unverifiable and therefore unscientific.

In the Vienna Circle we are accustomed to describe such statements as nonsense." But in recent philosophical work the boundary been science and values has been found more porous. So let's explore more closely the relationship between science and moral values. There are two potential directions of influence, so we should consider them in turn.

Values clearly have an impact on science. Thus, there are certain values even the most logically oriented philosophers would admit as central to the enterprise of science, namely the values of truth, objectivity, and empirical adequacy. These are acknowledged to be *good* features of a theory since they contribute to knowledge of the world. The decision to do science in the first place derives from the value of understanding and knowledge. More controversial is the role of non-epistemic values (those not related to knowledge), such as justice and human welfare. Involving such values in science is controversial precisely because they might conflict with the more fundamental epistemic values of truth and objectivity. Suppose a developmental biologist is ethically committed to the virtues of individual responsibility and self-determination. If such values enter directly into his scientific practice, then he may be inclined to minimize the evidence for various kinds of genetic determinism. The good, as he sees it, will have triumphed over the true. Nevertheless, as will shall see, a strong case can be made that the complete elimination of values from science is not only unrealistic, but also counterproductive: sometimes pursuit of the good is essential or conducive to discovery of the true.

As we saw in the discussion of gender, values can enter into science in at least three ways: the selection and prioritization of research, the practice and methods of science, and the content of science. Concerning the first, it seems clear that non-epistemic values do (and should) play an important role. This is perhaps most evident in medical science. The reason so much more funding is devoted to cancer research than to acne research is that cancer is much worse than acne. Values also enter into the determination of research priorities in more esoteric, or "pure," areas of science. Values and interests help to explain why the complete human genome was sequenced before the dog genome, and perhaps why the cat genome was sequenced later still (it pays to be man's best friend). Currently, large resources in chemistry and physics are devoted to areas of research related to climate change and alternative energies. Even in cosmology and particle physics, research is favored which promises "spin-off" applications relevant to space exploration or computing technology. In 1993, the U.S. Congress cancelled a multi-billion dollar "super-conducting" particle accelerator planned for Texas. Preference was given to NASA's international space station project, partly due to public perception that the latter offered greater long-term economic benefits. Given the amount of public expense involved in cutting-edge research, the potential benefits of such research must be weighed against alternative public investments.

THE SPIN-OFF VALUE OF PURE SCIENCE: BACK TO THALES

Appealing to non-epistemic values, especially human goods, to justify research whose value is otherwise mostly epistemic is nothing new. Aristotle tells us that the very first cosmologist, Thales of Miletus, made a fortune cornering the olive market. His investments were based on weather predictions derived from his natural philosophy. Thales' goal was not wealth (though presumably that didn't hurt) but to demonstrate the practical value of theoretical wisdom.

Even in a perfect world, with unlimited resources for "pure" science, we might want to block some avenues of inquiry because their social implications are bad. Consider an especially dramatic example from contemporary science. Some particle physicists have raised the possibility that the new large hadron collider (LHC) near Geneva could produce a "microscopic black hole" that would swallow the earth. Representatives of the LHC recently re-asserted the apparent scientific consensus that the risk is negligible based on the fundamental physics governing the experiments. They also point out that the same high-energy conditions produced in the LHC have probably occurred naturally millions of times in the history of the planet. We will discuss in detail the difficulty of assessing such risks in the next chapter. Still, it seems clear that if the black hole risk were real then the non-epistemic value of life on earth would trump the value of the theoretical knowledge promised by LHC. At the very least, as legal scholar Richard Posner has observed in discussion of similar risks from the upgrade to the Relativistic Heavy Ion Collider (RHIC) in Brookhaven, NY, "One hopes that before a decision is made on the funding request the proposed upgrade will be subjected to a careful cost-benefit analysis by neutral experts."

Notice that in evaluating the LHC risks, as well as the ethics of medical research involving human and animal subjects, the tension is typically between moral values and certain *means of obtaining* knowledge. It is the potential harm of the experiments, not the knowledge they provide, that seems morally problematic. But perhaps some knowledge is harmful or morally problematic in itself; maybe some things just shouldn't be known. Consider highly incendiary issues like the relationship between general intelligence or IQ and gender or race. Given the history of sexual and racial discrimination in most parts of the world, and the unsettled state of genetic explanations of complex traits like IQ, investigation of genetic sources of IQ variation may serve only to reinforce prejudice. Perhaps science shouldn't go there.

A possible retort is that this research may ultimately reveal no link between race or gender and IQ. Even if there is a link, this may be a good thing to know for guiding social policy, or "improving man's estate" as Bacon put it. Critics of IQ, such as the biologist Stephen Jay Gould, have responded to such arguments that IQ is really just a social construction measuring the aptitudes possessed, whether from nature or nurture, by the dominant class. So an uneven IQ distribution would reveal nothing more than social inequality. This is a matter for ongoing debate. But the debate is itself a clear example of the influence of moral and political values on the direction of science.

Aside from the selection and funding of research programs, non-epistemic values also influence the methods, and even the content, of science. The results can be disastrous from a strictly scientific point of view. However, social and political influence can also impede the growth of scientific knowledge. For example, in the mid-twentieth century in the Soviet Union, Stalin's agriculture minister Trofim Lysenko orchestrated a vigorous campaign against genetic accounts of evolution, proffering instead a version of Lamarck's "inheritance of acquired characteristics." Lysenko denounced genetics as "bourgeois" and "reductionist" and championed his own bio-logical approach as "practical" and "progressive." To its consider-able detriment, genetics was systematically purged from Russian science and education until the 1960s, when Lysenko himself was denounced and modern biological ideas were reintroduced. As I mentioned briefly above, something similar occurred with physics during the Nazi era. In the early part of the century, conservative and nationalist factions within German science (also present in French and British science) exploited the rise of Nazism to oppose Einstein's theory of relativity and, to a lesser extent, the quantum theory expounded by Heisenberg. The classically oriented, deterministic "German physics" advocated by Philipp Lenard and Johannes Stark was elevated above the "Jewish physics" of relativity and quantum mechan-ics. The so-called German Physics movement set back German science less than Lysenkoism did Russian science, but still illustrates the pernicious influence that political power can exert on science.

Science in modern liberal democracies is not immune from such influence. With so much medical research dependent on the pharmaceutical industry, researchers have strong vested interests in minimizing or suppressing evidence of ineffective drugs and adverse side effects. Some have argued that the Bush Administration's advocacy against models of climate change, Darwinian evolution, and embryonic stem cell therapy, three areas of comparative scientific consensus, represented an effort by religious and corporate interests to manipulate science. The Administration's frequent insistence on the "uncertain" and "tentative" status of current science, especially evolution and climate change, is somewhat reminiscent of the Vatican's stance during the Galileo affair that Copernican astronomy could be affirmed "hypothetically, but not absolutely."

Is the intrusion of moral and political values into the practice of science always detrimental to science? Shapin and Schaffer's study of the Boyle-Hobbes controversy seems a reasonably clear case of political ideals molding scientific opinion. But this influence did not hinder the advance of seventeenth-century science. If anything, the stark ideological confrontation between Boyle and Hobbes helped clarify the nature of the experimental method and the boundary between science and philosophy. Another episode mentioned in chapter one provides an interesting illustration of how external, non-scientific factors can promote the traditional aims of science. In 1277 the Archbishop of Paris, Etienne Tempier, condemned numerous widely held scientific doctrines, ostensibly because they imposed limits on God's power to act in nature. By placing God's absolute power above the authority of Aristotle, Tempier's decrees encouraged natural philosophers to think "outside the box." Historians believe this was a crucial factor in the gradual turning away from Aristotelian science that culminated in the Scientific Revolution.

Scientific progress can also be stimulated by the injection of non-epistemic values from within, by scientists themselves. Evelyn Fox Keller's work suggests that Barbara McClintock's discoveries in gene regulation were enabled by her compassionate and holistic value system. More recently, Alison Wylie and Lynn Hankinson Nelson have shown how the dramatic increase of women in the field of archeology since the 1970s fostered a greater sensitivity to gender in the study of early humans. The new generation of women archeologists perceived important information in the historical record that was invisible or irrelevant to earlier researchers. For example, faced with a puzzling disproportion of male skeletal remains in certain Australian Aboriginal villages, researchers initially proposed various ad hoc explanations, like divergent burial practices for the two sexes. It turned out that the sexual classification of the skeletons, which generated the original anomaly, was based on a scale of overall "robustness" that assumed early women were a much "weaker sex" than men. According to Wylie and Hankinson Nelson, recent, feminist-oriented research has corrected the underestimation of the size and strength of female humans and therefore solved this puzzle of modern archeology. Other studies with an emphasis on female agency have uncovered the likely role of early women in hunting activities previously reserved for males, thereby blurring the time-honored hunter/gatherer dichotomy. It seems that whether the application of values in science hinders or advances its fundamental epistemic goals will heavily depend on the values in question. While the values and political aims of Archbishop Tempier, Boyle's Royal Society, and the feminist archeologists are very different, they all encouraged alternatives to entrenched perspectives. In this they differ from the Stalinist central planning favored by Lysenko, the racist mythology of Nazi science, and the biblical literalism of many American creationists. Values contribute to science not because of their content but because they can spur the imagination and loosen dogmatic thinking. But if values become unresponsive to criticism they can foster distortions of their own.

So much for the impact of values on science. Does science also affect values? It is clear enough that modern science has momentous moral implications. The advance of genetic science, for example, introduces pressing moral concerns about the control of genetic information and about the genetic "enhancement" of children. As important as they are, these problems concern the difficulty of applying existing moral principles to emerging technologies, rather than the influence of science on morality itself. Thus, the issue of genetic enhancement involves an apparent conflict between the traditional values of parental rights and inherent human dignity. What I would like to consider, however, is whether science can be a source of morality itself: can learning about nature change our understanding of right and wrong?

The great seventeenth-century philosopher Benedictus de Spinoza, writing a hundred years before Hume, believed the new science revealed the subjectivity of traditional value concepts like goodness and perfection, evil and corruption. These are mere psychological projections of our own desires and aversions, which in turn depend on peculiarities of our brains. This explains the saying, Spinoza observes, that "brains are as different as tastes." Since moral concepts are mere projections, or "beings of the imagination" as he put it, we should judge things strictly in terms of their intrinsic nature and causal powers (as divulged by science): "the perfection of things is to be quantified solely on the basis of their nature and power; and consequently things are not more or less perfect because they delight or offend human sensibility." The vision implicit in Spinoza's analysis of moral concepts is that hard-nosed psychology and brain science will explain away traditional morality and replace it with a more realistic conception of our natures and hence of our perfections (and imperfections).

Psychology and brain science have advanced greatly since Spinoza's time, especially with the development of highly precise imaging techniques (e.g. fMRI) in the last few decades. Has Spinoza's prediction of a radical undermining of traditional moral notions been confirmed? Consider the principle of retribution, which says that wrongdoers should be punished simply because they deserve to be. The principle is an ancient one – think of the Old Testament maxim "an eye for an eye; a tooth for a tooth" – which answers to our moral sense that humans should be held responsible for their actions. I say "humans," but in the Middle Ages it was not uncommon to punish or execute animals, such as pigs and dogs, for assault and theft. The aim was not simply to eliminate pests, for the animals were subjected to elaborate trials aimed at achieving justice. The practice seems bizarre to us now because we do not conceive of animals as morally responsible in the way humans are. Part of the reason for excusing animals is that we have learned from the study of animal behavior and physiology that they have limited power to control, for example, the "fight or flight" response.

But suppose scientific study leads us to conclude that many humans are similarly unable to control their violent behavior. This is a major issue in the emerging field of "neuro-law" which investigates the neurophysiological basis for criminal behavior. A 2005 U.S. Supreme Court ruling against execution of minors relied in part on adolescents' "comparative lack of control." An amicus brief provided by the American Medical Association (AMA) argued that since "adolescent brains are more active in regions related to aggression, anger, and fear, and less active in regions related to impulse control" to punish them for capital crimes is to "hold them accountable not just for their acts, but also for the immaturity of their neural anatomy." Other investigations have found common patterns of imbalances in neurotransmitters and hormones among violent criminals. If neuroscience and neuro-law show that some behaviors for which we hold people responsible, both despicable and heroic, are largely a consequence of brain processes over which they have no control, we may need to rethink the principle of retribution itself. It may still make sense to incarcerate criminals for the threat they pose to the rest of us, just as we sometimes quarantine the sick. But not because they *deserve* such treatment.

The AMA's argument against the execution of youth seems to turn on the unimpeachable premise that we shouldn't be held responsible for things we can't control. As philosophers put the point: "ought implies can." But the social sciences may reveal that much of what traditional morality demands of us presupposes on an unrealistic conception of what we're capable of. For example, numerous psychological studies indicate that our tendency to act virtuously or not is strongly affected by factors that ought to be irrelevant from a moral point of view. One study found that we're much more likely to help someone who has just dropped their bags if we've recently found a small amount of cash (an increase from only four percent helping to eighty-five percent with the found change). Other studies have shown that generosity varies greatly with factors we aren't conscious of, like ambient noise and surrounding odors (apparently the entrances to bakeries make a good location to panhandle).

The moralist will of course reply that nothing follows about how we *ought* to behave from the rather grim facts about how we *do* behave. Yet, surely what we morally demand of people should be sensitive to the sorts of beings they are. That's one of the reasons we expect less, morally speaking, from small children and animals. As the social sciences, combined with evolutionary biology and the neurosciences, map out the deep predispositions of human nature, we may come to hold ourselves to a lesser, or different, moral standard than the classical moral theories have done. In the meantime, we are stuck with the problem of how to live. We will be able to address this problem sensibly only if we take careful account of the options and challenges for human existence that future science is likely to provide. This is the subject of our final chapter.

6
Science and human futures

We began looking back to the origins of science. We now look forward to where science and technology may take us. For better or worse, the futures of science and humanity are linked. The pace of technological change is certainly accelerating in areas that are certain to transform human existence, such as computing, genetic engineering, and nanoscience. Some believe that we are close to a "tipping point" or "singularity" when technologies like artificial intelligence become advanced enough to begin improving themselves. The future is hard to predict, especially if we entertain runaway progress of this sort. We should be skeptical about exotic prophecies and doom-sayings, as countless failed millennial predictions testify (the most recent prominent example is the "Y2K" panic). As David Hume warned in his discussion of miracles, we need to scrutinize such reports carefully since humans take pleasure in "wonder" and are apt to be dazzled and defrauded. Futurism has a long history of economic, but not necessarily predictive, success. That said, we certainly can anticipate the future and to some extent control it. We know the trajectory of scientific progress in the last few hundred years and the uses to which humans have so far applied the undeniable advances in knowledge. Based on this we should consider where science is likely to take us in the future, and how, given what's at stake, we might play a part in charting its course for the best.

Are we doomed?

The vast majority of species that have ever existed on Earth have become extinct. Why should we be any different? Well, for one, we are unique in having the capacity to reflect upon our collective prospects; it is this capacity itself that uniquely equips us for survival, since we can anticipate and respond to both immediate and long-term threats. Warnings of the apocalypse are as old and varied as creation myths. But just as science has given a more accurate picture of the remote past, it also offers some basis for evaluating our future. The news is not so good. A sober estimation of our prospects should cause real concern, if not outright pessimism. According to Nick Bostrom of the Oxford Future of Humanity Institute, "there seems to be a consensus among researchers who have seriously looked into the matter that there is a serious risk that humanity's journey will come to a premature end."

Let's survey the main risks to humanity.

Conflicts among nations, ethnic groups and regions are likely to increase with rising demand and diminishing supplies of fresh water, wild foods, oil and arable land. The targeting of civilian populations in wars or other political conflicts, which became common in the twentieth century, will also increase as "weapons of mass destruction" become cheaper, more effective and easily portable. Of particular concern is the large stockpile of nuclear warheads left over from the Cold War. Falling into the wrong hands, one of these could easily trigger a full-blown nuclear conflict in Central Europe, the Indian Subcontinent or the Middle East. A regional nuclear war could kill millions, but would be unlikely to lead to human extinction in the short term. The threat of global nuclear war remains and could rise to cold war levels if Russia reinvigorates or the United States continues to act aggressively. Any serious nuclear conflict would cool the climate drastically and damage the protective ozone layer.

Attacks involving biological agents like germs and viruses may be an even more serious threat given their accessibility and ease of delivery. More frightening still is the prospect of biological pathogens, genetically engineered with built-in resistance to all known antigens, that would kill humans directly or attack food and water supplies. Such a weapon might be strengthened by nanotechnology, incorporating molecular-scale machines to increase the potency, durability and reproductive capacity of the parasite. Whatever the weapons employed, any war or large-scale terrorist attack could precipitate a breakdown in the global economy and more wars during the ensuing panic, struggle and sickness.

After war, the next greatest threat to humanity is from disease. Many experts think a global flu pandemic, most likely originating in Asia from a virus that "jumps" from domestic fowl and mutates quickly, is inevitable in the coming years. Although there have been global flu pandemics in the past, notably the 1918 "Spanish flu" which killed upwards of 50 million people, a new pandemic could be much worse owing to increased population density and air travel. Besides the massive deaths from the original virus, an even greater risk comes from the threat to regional infrastructure and security that would accompany wide-scale sickness and economic collapse. Just as war can foster disease (the Spanish flu followed on the heels of World War I) the reverse can also occur. In the aftermath of a pandemic, which would probably hit already impoverished regions the hardest, scarcity could feed civil strife, violent extremism, and spiraling chaos.

Humanity may be at risk from an accidental technological catastrophe. Systems of communication and energy transmission have become so large and interdependent that it is difficult to predict or control breakdowns. A massive failure of the internet would cripple infrastructure worldwide and set-off a financial panic, with unpredictable consequences. Another risk is that research in genetic engineering could inadvertently produce and release a vaccine-resistant strain of a deadly virus like bubonic or black plague. In 2001 two medical researchers in Australia genetically engineered a new strain of "mouse-pox" that was resistant to all vaccines. A researcher, by accident or by malicious intent, could conceivably do the same with the human smallpox virus that killed millions before vaccination became widespread in the twentieth century.

As nanotechnology continues to be improved, it is possible that a hyper-replicating "nanobot" could be released into the natural environment, creating a "grey goo" that exploits almost any source of energy and eliminates all its natural competitors before turning on itself – the mother of all murder-suicides. Even more exotic, and correspondingly more catastrophic, accidents could result from the high-energy collisions that will be routinely produced in the new generation of particle accelerators. In the previous chapter I mentioned the risk of an accelerator-induced "microscopic black hole." Another possibility, entertained by the philosopher John Leslie, is that our universe is merely a "meta-stable" vacuum, i.e., there is a lower energy state into which it may collapse. If this should happen as the result of a particle accelerator collision, the energy released would destroy us all in an instant. Physicists have promoted the new LHC near Geneva as approximating the conditions shortly after the big bang. But doesn't that mean that there is a risk of inadvertently producing "big bang v. 2.0?" Most experts agree that all these scenarios are ruled out by currently accepted physical principles. Then again, the experiments are being conducted precisely because we don't fully understand the most basic physical principles.

And of course there is climate change. Most serious analyses of the likely impacts of climate change on humans (assuming no significant, coordinated reduction in carbon emissions) predict a significant and sustained downturn in the global economy, increased hunger and malnutrition, a large reduction in global bio-diversity, more weather-related disasters, permanent loss of farmland, fresh water, fish stocks, and many ocean-side communities. There are also various, somewhat more speculative, feedback scenarios in which the initial effects of temperature increases trigger processes, like the release of methane from melting permafrost, that combine to produce a "snowball" of warming. Global warming obviously warrants serious international action, but it is unlikely to produce massive death or suffering. Even in the grimmest of the six warming scenarios evaluated by the recent U.N. Intergovernmental Panel on Climate Change predicts only a 26–59 centimeter rise in sea levels by the end of the twenty-first century. As with pandemics, the most serious risks to humanity are from social side effects like wars and terrorism resulting from the disproportionate impacts of increased water and food scarcity upon underdeveloped regions.

The good news is that because all of these risks are products of human technology we have a reasonably strong understanding of their internal and social dynamics. And this might enable us to prevent them. For example, large investment by developed nations in underdeveloped and third world nations could reduce the appeal of religious and sectarian fanatics, a major source of terrorism. International agreements regulating or banning dangerous biotechnology and nanotechnology could be implemented. All nations could adopt the Kyoto Protocol and other U.N. targets and conventions on climate change. Using science and diplomacy, we can also prepare for disasters that we have done nothing to encourage. For example, even if we are on a collision course with a giant asteroid, such as many scientists believe killed off the dinosaurs, we can develop an international plan to identify and deflect it before it's too late.

A hopeful example of using international cooperation to avert global catastrophe is the effort made to slow and reverse the depletion of the Earth's ozone layer, which helps to filter harmful UV radiation. Thanks to a 1987 agreement (the "Montreal Protocol") limiting the production and marketing of CFC's (a major cause of ozone depletion) in consumer products, the ozone layer now seems to be recovering. Are we likely to achieve similar successes overcoming the long-term threats described above?

Deciding how best to act in the face of possible danger requires a clear-headed evaluation of at least two factors: (i) the *costs* of acting to avoid the danger or not; (ii) the *probability* of avoiding the danger if we act or not. Combined, these two factors give us what decision theorists call the "expected costs" of a given course of action or inaction in a given set of circumstances. We rely on calculations of expected cost every day, at least tacitly. Suppose I am late for an appointment and consider flagging a taxi. I consider the cost of the taxi, the cost of missing the meeting and the probability that I will make it by cab vs. walking (or running). Similarly, in deciding how to act in the face of climate change, we need to consider the cost of reducing carbon emissions, the cost of a continuing rise in temperatures, and the probability that the temperature rise will be slowed if we act or not. Mathematically, we can think of the expected cost of an action or inaction as the product of the expense of the outcome and its probability, summed across each of the possible outcomes. For example, consider someone who is offered flood insurance on a million dollar home for a $5000 annual premium. If the risk of a home-wrecking flood in a given year is one percent then they ought to buy the insurance, since the odds of a flood are small but the loss of the house is a very great cost. The situation is represented in the following matrix, with the possible actions on the left and the possible outcomes on top.

Table 6.1

	No Flood	Flood	Expected Cost
Not Insure	$0 \times .99 = 0$	$1,000,000 \times .01 = 10,000$	10,000
Insure	$5,000 \times .99 = 4950$	$5,000 \times .01 = 50$	5,000

The choice is easy since the expected cost of buying the insurance is one half of not doing so.

This approach to risk is pretty straightforward and seems to tell us that we ought to act quickly and decisively to avoid the dangers to humanity listed above, since the costs of inaction are so great compared with the costs of action. If the cost of the taxi is much less than the cost of missing the meeting, and it's very likely that I'll miss the meeting unless I take the taxi, it's clear what I ought to do. Likewise, in each of the cases above, the cost of acting is real, but nothing like the cost of not acting. Furthermore, although there is a chance the catastrophes will not occur if we don't act, or occur even if we do, the costs associated with these scenarios are still dwarfed by the costs we risk by not acting. For example, even if there is only a small chance that self-replicating "nanobots" will fall into the hands of terrorists if we don't ban this technology, and even if there remains a chance that a ban won't succeed in foiling terrorists, the catastrophic consequences of nano-terrorism mean we should expend significant costs to foil such plans now. It's as if the taxi was our best chance at making our one and only shot at the job of a lifetime.

This is not to say that *any* immediate costs are warranted in the face of *all* remote risks. Even on the relatively high ground that I happen to occupy there exists a small chance my home will be flooded; but I don't bother with flood insurance. I would hate to lose my home but I would not pay even $100 a year to insure against flooding (or meteor impact). Some employers provide "accidental dismemberment" insurance at very low cost to their employees. The price is so low because in most jobs the chance of losing a limb is practically nil.

Still, some people are risk averse: "better safe than sorry." They are willing to spend more now to avoid a very bad outcome even when the risk is minute and the short-term costs are significant. For example, in the insurance scenario above, the risk averse might be willing to spend much more than $10,000 against the expected cost of a flood, and perhaps take other measures as well, to prevent the loss of their home. The Large Hadron Collider was started up despite the very small risks of a cosmic disaster. But an extremely risk-averse approach to technological catastrophe would require a ban on high-energy particle accelerators to forestall even the slightest chance of producing a not-so-microscopic black hole. In modern policy-making, aversion to risking disasters is captured by the "Precautionary Principle," which says that an action should not be undertaken if there is an uncertain risk of serious and irreversible harm. The Precautionary Principle would require even more vigilance about the risks we've mentioned than ordinary cost analysis.

But we are likely to adjust our behavior in response to these powerful considerations only if we both understand the enormity of the costs and also care about them. The problem is that the major risks are either remote in probability terms or remote in time. For example, even according to relatively dire scenarios, the major effects on humans from climate change are likely to be quite gradual. But humans are terrible at taking proper account of long-term risks, especially when those risks are uncertain. This bias in favor of imminent threat may have been written into our genes by evolution. In the savannah, a large mammal that forgoes the immediate benefits of available food in order to set aside reserves will likely fall victim to short-term thinkers. This preference for short-term gain might explain in part the persistence of obesity in prosperous Western nations despite the well-known individual and collective health costs over the long run.

Most of the threats to humanity don't pose much danger at all to those now in a position to act – only to "future generations." This raises a serious moral question: why should we care about future generations since they don't exist, at least not yet? Perhaps we have an obligation to those who are *likely* to exist not to despoil the environment and make their lives miserable. It would be wrong, by analogy, to plant a bomb with a 100-year fuse in Times Square or Piccadilly Circus, making the excuse that no one who now exists will be harmed. Since they will exist, we mustn't do what will harm them. But if humans do go extinct from war or disease then "they," those future generations, won't ever exist. Why should we try to prevent something for the sake of people who, if we fail to prevent it, won't exist in order to be harmed? It's hard to comprehend an obligation both not to harm future humans and also to ensure that they will exist. For if future generations can demand existence, as well as a healthy environment, it would seem to follow that healthy couples are now obligated to have children – many, many children – that they would otherwise have preferred to avoid through contraception.

As it happens, many people do desire to have children, not out of obligation, but by natural inclination. They also instinctively hope that their children will have children and that the needs of their grandchildren will be met. And for most of us this well-wishing extends to nieces, nephews, children of our friends, and so on. This strong and nearly universal concern for at least two "future generations," no doubt also evolutionarily conditioned, counterbalances our tendency to neglect long-term risks. So assuming that scientists and policy-makers are able to convince the public and their political representatives of the risks to future humans, there is at least a fighting chance the present generation will make the necessary sacrifices, if only for the sake of their own grandchildren.

THE DOOMSDAY ARGUMENT

A philosophical proof of doomsday:

1. Unless the human race goes extinct quite soon then you are among the very earliest of all human to have existed.
2. It is statistically very unlikely that you are among the very earliest of humans to have existed.
3. So, the human race is likely to go extinct quite soon.

The first premise is based on the geometrical rate at which the human species has grown (doubling in size over shorter and shorter spans of time). If this goes on for more than another 100 years or so – or even if the rate of growth stabilizes – then there will be many, many more people born later than you rather than earlier. The second, more philosophical, premise is based on the principle that all else being equal we should infer that we are in collections that would make our position in that collection typical rather than highly unusual. Suppose your friend enters your name in a raffle (assume only one entry per person), but neglects to tell you whether you were entered in raffle A or B. You know that 1,000,000 people entered raffle A, but only 10 entered raffle B. In each lottery, all the tickets are drawn but only the first five drawn win a prize. You receive a phone call informing you that you were drawn fifth and have won a prize; but in your excitement you forget to ask whether it was raffle A or raffle B. There is no need to call back. It is much more likely that you would have been drawn fifth from lottery B than from lottery A, since otherwise you were among "the very earliest to have been drawn." If this seems plausible, the doomsday argument should also.

But before you make changes to your will, consider these responses. Opponents have noted that the argument would have applied to every previous generation, back to the earliest humans. But every one of them would have drawn the wrong conclusion since humanity has survived. Another problem is that it is hard to apply the doomsday reasoning to the possibility, which can't be ruled out completely, that humans go on forever, perhaps by

THE DOOMSDAY ARGUMENT (cont.)

colonizing the galaxy. For what could it mean to be "typical" or "improbably early" in a collection that is infinite? It may also be possible to turn the logic of the doomsday argument toward a more hopeful conclusion. A baby born today will be among the very *latest* born of all humans if the race goes extinct very shortly after his/her birth. Since being born this late is extremely unlikely, the race will probably not go extinct very soon (so long as babies are being born). A related point is to apply the argument not simply to possible human histories (long and short) but to possible other civilizations (long and short) including alien ones. Since most intelligent beings are in long-lived civilizations, given long-lived civilizations have more members, we should assume we are also. So doomsday is probably far off. Of course, this last objection depends on the existence of other civilizations, which we will discuss below.

Avoiding the main risks to humanity requires collective action, both within and among nations. Democratic nations will be unable to enact environmental policies or disaster avoidance plans without the consent of the citizenry, either directly through voluntary participation, or indirectly through electoral support for coercive policies. But, when I consider my individual expected costs, a serious problem emerges about the rationality of my involvement in collective action (even apart from the problem of short-term bias). Consider a program of mass prophylactic vaccination against various avian flu strains. The point of such a program is not to protect individuals from an existing outbreak, but to introduce sufficient immunity in the population that a future pandemic cannot "get hold" when it emerges. If the program is voluntary, should I choose to be vaccinated? On the one hand, if others don't choose to vaccinate, then I no longer have a reason to do so: if there is going to be a pandemic my personal vaccination won't prevent it. I might have a reason to vaccinate if a pandemic does emerge, but by that time it will be too late to stop it taking hold on a mass scale, which was the point of the program. On the other hand, if most everyone else accepts precautionary vaccination then again I shouldn't bother for the same reason as before: my susceptibility won't make any difference with respect to the pandemic's taking hold. I can "free-ride" on the protection the others have afforded. Given the inconvenience, expense and side effects, my expected cost of being vaccinated is greater whether or not others choose to be.

The problem is that everyone else seems to be in the same situation as me. So, if we all act according to our individual self-interest, as given by the calculation of expected cost, none of us will act to avert the catastrophe. This ancient problem about the rationality of cooperation – which applies also to more familiar activities like recycling, polluting, voting, use of public transport, etc. – is known as the "tragedy of the commons." The name derives from the case of individual herders who each over-graze a common feeding ground leading to the loss of the resource for all.

For many problems the tragedy is forestalled by the existence of governments and their law enforcement agencies. Governments enable us to "bind ourselves" and our fellow citizens to rules which will sustain common resources, such as enforceable limits on fishing in public waters. One can still take a chance being a "free-rider" and over-fishing; but this will not reduce your expected cost since there is a real risk of a monetary fine. Similarly, many governments require childhood immunization against measles and other communicable diseases precisely because it may not be in the individual interest of parents to acquiesce in a program of voluntary vaccination. Coercive vaccinations of this sort may become necessary to prevent a flu pandemic.

The problem is that the policies of one government can't prevent a dangerous pandemic from taking hold in another country, mutating, and then spreading around the world. Most of the threats we face as a species cannot be solved only at the local or national level. Since the risks transcend national boundaries, effective preventative measures will require a general "buy-in" by the international community. It does no good for me to refrain from using lawn fertilizers that contaminate the local water supply if my neighbors refuse to follow suit, and it does no good for China to reduce carbon emissions if the other industrial and developing nations fail to comply. But here we find no larger authority, no enforcement agency above the sovereign nations themselves that would make it rational for each nation to fall in line. In the international context, the self-defeating logic of the free-rider kicks in with a veangeance. Without a world government, it's not clear how to avoid an international tragedy of the commons.

But in spite of the rational barriers to cooperation, many large-scale societies with generally benevolent governments have existed for centuries. And many of these, like the United States and Germany, were formed from smaller provinces or principalities. So there is reason to hope that substantial international cooperation can also emerge in time. The U.N. has already achieved important agreements on issues like global warming, war crimes, and pandemics. One reason for optimism is that humans seem naturally predisposed to cooperate. This might be surprising given our evolutionary background. Shouldn't natural selection favor purely selfish behavior? Consider a population of monkeys plagued by a deadly parasite that implants itself in places like the scalp that are hard for the infected party to inspect or groom. But they can be easily spotted and removed by others. Suppose there are two sorts of genetically based behavioral predispositions in the population: SUCKERS, who groom others even if they are not groomed, and CHEATS, who accept grooming but never groom others.

It should be obvious that the SUCKER gene will disappear quite quickly from the population. Even a population dominated by SUCKERS will be decimated in the long run if CHEATS occasionally appear by random variation. But it's equally clear that once the SUCKERS are gone, the CHEATS will be in trouble since there's no one left to remove their parasites.

So a species that benefits from cooperation, as we apparently have, must be neither pure SUCKERS (altruists) nor pure CHEATS (egoists). Consider a third variation: the GRUDGER will groom others but only if the groomed reciprocates. If there is a sufficient number of GRUDGERS, they will prevail over the CHEATS since they can rely on one another for grooming while the CHEATS can expect to be "free-groomed" only once before succumbing to the parasite. In this way, GRUDGEY behavior, which goes by the technical name of "reciprocal altruism," might be favored by evolution, at least for relatively social and intelligent species like us. Indeed there has been quite a lot of psychological and evolutionary confirmation of this hypothesis.

To see more clearly how our GRUDGEY nature might help us avoid the tragedy of the commons, consider a classic puzzle of decision theory known as the Prisoner's Dilemma. Suppose you and your partner in crime have been nabbed near a bank you just robbed. You've stashed the loot but the police have enough evidence to hold you for a few days in separate cells. A clever prosecutor offers each of you the following deal (and you know your partner has the same offer): if you confess and your partner keeps quiet, you go free and he gets twenty-five years. Vice versa if he confesses and you keep quiet. If you rat each other out, it's ten years each. Finally, the prosecutor concedes that he can only put you away for one year each if you both keep quiet. But with a knowing smile he says he's pretty sure that won't happen.

Table 6.2

	Partner CONFESSES	Partner DOESN'T CONFESS
You CONFESS	Y: 10 years	Y: 0 years
	P: 10 years	P: 25 years
You DON'T CONFESS	Y: 25 years	Y: 1 year
	P: 0 years	P: 1 year

For he knows that each of you is looking out only for himself. He also knows you're both pretty smart, and so he expects that you'll each reason as follows:

If he confesses (the rat!), then I get twenty-five years if I keep quiet but only ten if I confess too. If he keeps quiet (the sucker!), then I can sit in jail for a year by also keeping quiet, or go scot-free by confessing. So whether he confesses or not, I'm better off confessing!

The problem is the same reasoning applies to your partner, and so you both end up confessing and receiving ten-year sentences even though you could have had only one year by simply keeping quiet.

Perhaps you might keep quiet out of loyalty to your gang. But remember: we're trying to explain why such cooperative dispositions would exist in the first place. More likely, you might each cooperate out of fear that the other will later retaliate somehow. Suppose the two of you have been through this several times before (maybe with shorter sentences at stake). In one of the earlier dilemmas, you kept quiet as a signal that you were willing to cooperate. In response your partner kept quiet the next time, and a mutually beneficial pattern of cooperation was established. With all the extra time on the street, the two of you train the younger gang members to always be loyal, never trust the cops, fear ruthless retaliation, and so on. In the long run, your gang spends less and less time in the slammer.

This idea that a long run of iterated prisoner's dilemmas might favor reciprocal altruism, or what decision theorists call the "tit-for-tat" strategy, has been proposed as an abstract model for how cooperation actually evolved. It has also been applied to real international dilemmas, such as arms races and environmental treaties. If we can exploit our evolutionary propensity to "take a hit" in order to achieve long-run benefits, perhaps with a small gesture of unilateral disarmament or voluntary CO_2 reductions, there might yet be a happy ending to our global tragedy of the commons.

Other worlds

If the world ends not with a bang, but a whimper – slowly becomes uninhabitable, depleted of natural resources, or over-populated, humanity may be able to continue on other planets. Even if things go well for us here, it might make sense to colonize space so that Earth can be repopulated in the event of an unforeseen catastrophe. Humans have already spent brief periods on the Moon and now sustain a continuously manned space station. The most attractive locations for permanent human outposts in the short-term are the Moon and Mars, or perhaps a large asteroid. There are serious technological barriers that will need to be overcome to make colonization of the Solar System feasible. Using even the fastest means of space transport available, a return trip to Mars would take at least a year. Travelers could experience significant bone and muscle loss owing to reduced gravity in transit. This is especially serious given the need to rapidly decelerate or "brake" on the approach to Mars. A reliable means of simulating gravity in flight, possibly the centrifugal force in a spinning craft, along with a rigorous exercise program, could ameliorate this problem. But the flight would face external dangers too: asteroid collisions and cosmic radiation, for example.

Given the astronomical costs of even a single supply trip, a colony on Mars would need to become self-sustaining in short order. So means of producing energy (presumably nuclear), oxygen, and food (presumably plants grown in greenhouses) would need to be developed. Water supply would be plentiful, in principle, since there are polar ice caps and probably underground aquifers on Mars. Since the atmosphere is mostly carbon dioxide and the temperature is usually below 0°C, life will be mostly indoors. After the first few return trips, all those departing for Mars would be leaving for good. Despite these challenges, manned trips to Mars have been urged by eminent astronomers like Carl Sagan and Stephen Hawking, and are part of NASA's long-range plans.

Along with the other planets, eventually other stars (those with planets approximating ours) would need to be colonized since our own sun will begin to burn out in 5 billion years – long after it engulfs the inner regions of the Solar System. Whereas interplanetary travel takes months and years, an interstellar journey would take decades or centuries. And the barriers to such journeys are not merely technological, but physical, since according to the theory of relativity nothing can exceed the speed of light. Therefore, all of the difficulties of manned interplanetary travel and colonization would be multiplied and supplemented with the problem of journeys that take longer than the lives of the passengers. Some form of suspended animation, as envisioned in science fiction such as Stanley Kubrick's film *2001: A Space Odyssey*, would be employed, or perhaps long hibernation combined with ordinary reproduction. Natural birth and death were not uncommon experiences in the long sea journeys made by the Earth-bound colonists of the sixteenth century; many interstellar colonists would spend their entire natural lives in passage. Given the uncertainty of the conditions upon arrival, and the sheer length of the journey, the ship would need to be already an essentially self-sustaining "world-vessel" upon launch: the large community of voyagers might even enter and become accustomed to its conditions years before departure, and not venture out until years after arrival upon an alien planet. Return to the Earth would be out of the question and even communication would be very tenuous considering that messages would be received, at the earliest, about eight years after being sent.

Proposals have been made for less expensive means of interstellar transport. Since the main drags on large spacecraft are the fuel, cargo, and passengers it carries, the smaller the better for long journeys. Nanotechnology might allow very tiny craft to travel at high speeds with comparatively light fuel demands. They could not carry our bodies, of course, but they could perhaps carry digitally encoded DNA or digital blueprints of our brains. So if the "nanobots" could rely on the resources of the target stars, and were self-replicating, then we might be able to transport computer "emulations" of ourselves to safe havens much more cheaply than by conventional bodily transport. Thus, in his book *The Physics of Immortality*, cosmologist Frank Tipler maintains that "the fundamental reason for allowing the creation of intelligent machines is that without their help the human race is doomed. With their help, we can and will survive forever."

Not so fast. Even assuming (what is hardly self-evident) that a computer emulation of my genetic or neurophysiological make-up would be genuinely conscious or intelligent, it is very unclear that if it survived then "I" would survive. For even if such a being seemed to itself to be a person who had previously lived on Earth, and was now settling into a new life in the neighborhood of Alpha Centauri, that being would seem to be a mere *simulation* of me. To see this, suppose such a nanobot *had already been* sent off at near-relativistic speed on your twelfth birthday, without your knowledge, and is now happily exploring and populating another planet. Presumably, you would not conclude that *you* are now on another planet. And so it is hard to see why it would give us any consolation if the Earth were about to be destroyed to know that computer emulations would be safely launched on their long journey to "our" resurrection in the new world.

Especially if we stick to ordinary bodily transport, the question of interstellar colonization is largely academic since the technological barriers and costs are so overwhelming. Still, it's worth reflecting on whether we really should *want* to colonize the Milky Way (and beyond!) even if we could. The putative motivation would be to save the human species in the event of catastrophes in our own Solar System. But since these colonies, especially the interstellar ones, would be reproductively isolated from us, they would begin to diverge genetically by natural evolution, perhaps speeded by genetic engineering. Within a few thousand years, even the relatively nearby colonies might seem like genetic strangers. In the long run, we will have populated the galaxy with species that are very different from one another and from ourselves. But why should this matter to *us*? Suppose we discovered conclusive archeological evidence that some ancient technologically savvy progenitors of humans had left the Earth for other galaxies 200,000 years ago. Would that give us comfort in the face of an approaching apocalypse?

Perhaps the point of colonization is not so much to spread Homo sapiens as to spread biological complexity and intelligence in an otherwise lifeless and unthinking universe. Assuming our remote descendants would be intelligent, is there any reason *why* – other than simple bio- or logo-centrism – the universe would be better off than it is without the colonizers? Some of the scenarios for colonizing Mars involve permanently "terra-forming" its atmosphere to make the planet friendlier to human life. This should bring to mind the extent to which indigenous cultures and landscapes were molded to suit colonizers from

Western Europe. Before undertaking such a transformation of other planets, we should think hard about whether this is within our rights and whether it would really be for the best.

Of course, Mars is probably just a cold and lifeless desert, and maybe no argument is needed for why a world with science (and love) is better than one without. But we need to bear in mind that human colonization on Earth also involved tremendous suffering and strife. Indeed, the human colonization we've experienced suggests a real risk of exploitation and rebellion. Do we really want to extend such behavior throughout the universe? A likely source of conflict between space colonies and the "old world" would be control and allocation of resources. In a sense, we already face this conflict since money now spent on space exploration technology, which will benefit only future generations, can't be used for present needs. Even as enthusiastic a supporter of Mars exploration as Carl Sagan concluded that in the short term "the most important step we can make towards Mars is significant improvement on Earth. Even modest improvements in the social, economic and political problems that our global civilization now faces could release enormous resources, both material and human, for other goals." As with the more mundane insurance against catastrophic risks, there is an inevitable trade-off in space exploration between present needs and future hopes. This raises one final concern about planning for departure: it may encourage a dangerous complacency about the challenges we face on Earth. If humanity came to regard Earth as a mere starting point, rather than our permanent home, we may be inclined to treat it the way hotel rooms and rental property are routinely mistreated.

If the ultimate point of colonization is to ensure the existence of complex biological organisms, consciousness, and intelligence, not just the human species, then we ought to consider whether these already exist elsewhere and possibly save ourselves some trouble. How likely is extraterrestrial intelligence (ET)? From what we know of cosmology, it seems highly likely. There has been sufficient time. The universe is twelve billion years old, but evolution on Earth required only about four billion years to produce humans from simple cells. There are sufficient incubation sites. There are, after all, hundreds of billions of galaxies in the universe each with billions of stars. Planets have been detected recently circling nearby stars, and there is no reason to think those stars are special. In our own solar system, we know that one planet has life, and two others (Mars and Venus) have conditions that could have permitted life in the past. Given the huge number of planets in the universe with temperatures compatible with life, and the vast amount of time available for evolution, it seems unbelievable that intelligent beings would appear only once. If we employ the principle used in the "Doomsday argument" above, that we should prefer hypotheses that don't make our circumstances extremely unusual, it seems to follow that ET is very likely.

Indeed, according to one prominent version of big bang cosmology, the so-called "inflationary model," because the universe is spatially infinite (in a technical sense that needn't concern us here) and uniform in the large (the laws of physics are the same everywhere), and given these laws allow only finitely many possible physical permutations, it follows that any physically possible arrangement that occurs anywhere, occurs at infinitely many places. Imagine observing a single massive roll of an infinite collection of dice. Since the dice are assumed fair, each of the results 1–6 is going to occur equally overall, i.e. an infinite number of times. Astonishing as this may be, it means that every cell, every genome, and every civilization, has infinite copies distributed throughout the universe as a whole. Obviously, there will then be other intelligent beings besides us. Still, even if this (admittedly controversial) argument is sound, it does not follow that there is intelligent life in any portion of the universe that we could be in physical contact with. Indeed, given what we know about the complexity of DNA, not to mention human brains, these portions would be so sparsely distributed as to be nearly as remote as fictional worlds. That is, they are so distant that light would not have time to reach us given the total age of the universe (beyond the "particle horizon" or "observable universe"). Paul Davies estimates that we would need to sample 10^{39943} galaxies to have a chance of finding duplicate DNA; but the observable universe is only about 10^{10} galaxies. So although some cosmological models seem to guarantee the existence of ET, they certainly should not lead us to expect encounters.

It may still seem reasonable to assume, given the sheer number of stars and planets in our own galaxy, that we are not alone. After all, we are intelligent and there's no reason to think we're atypical. Perhaps our own existence suggests there is something about our galaxy that favors life and intelligence and so we should be prepared for encounters. This sort of reasoning can be misleading however. If this *is* the only planet in our galaxy that permits intelligence, then our being on it is hardly surprising or atypical – where else *could* an intelligent species find itself but on a planet that supports life? In that sense, our being on an intelligence-permitting planet can't be coincidental, whether there is only one such planet or billions. To illustrate this point, consider a massive game of Russian Roulette, where each player has his own gun. It happens that the gun I fire is loaded with a blank. I would be hasty to conclude on this basis that most of the other shooters are also safe since "there's nothing special about me." For even if there is only one unloaded gun among thousands it's not a *coincidence* that the only survivor chose the only unloaded gun. How could it have been otherwise?

Such reasoning relies on what cosmologists call the "Anthropic Principle." Here's how an early formulator of the principle, Brandon Carter, puts it: "what we can expect to observe must be restricted by the conditions necessary for our presence as observers." In the Russian Roulette case, our having an unloaded gun is required by the observation that we are standing there puzzling about this, and so tells us nothing about the other guns. The upshot of anthropic reasoning for the question of ET is that we cannot regard the fact that we are on an intelligence-permitting planet as evidence there are many such planets because we will observe ourselves to be on an intelligence-bearing planet no matter how common or uncommon they are. Put differently, the hypothesis that we are alone and the hypothesis that we have neighbors equally predicts, and are equally confirmed by, the observation that we would find ourselves at home. In order to decide whether we are in fact unusual, other information besides our mere existence must be relied upon.

One important argument that we are probably alone takes all the same data the argument for ET used about the probable millions of life-permitting planets in our galaxy, and adds the pertinent bit of information that we haven't heard from anyone. Extending the Russian Roulette analogy, suppose at the end of the game everything is eerily, even deathly, quiet. This seems to show, perhaps surprisingly, that you are in a game with only one (unloaded) gun. There are about 100 billion stars in the Milky Way alone. If we are not alone, then one of the other civilizations ought to have reached a level of technological advancement that would enable them to contact us. The question, originally posed by the physicist Enrico Fermi, is why haven't we heard from them? The obvious answer seems to be: because they're not there.

The physicist Frank Drake tried to formalize Fermi's question to some extent by providing an equation that includes factors for the number of livable planets in the galaxy, the probability that life would emerge on such a planet, the probability that an intelligent civilization would develop means of interstellar communication, how long such a civilization was likely to last in order to be able to contact us, and so on. Plugging in different values for these variables produces wildly different results for the total number of communicating civilizations. (Drake himself recently upped his own original estimate to greater than 10,000). But, discounting the bizarre and mostly discredited "close encounters" reported in supermarket tabloids, it seems none have made contact. And it's not like we haven't been looking. The Search for Extraterrestrial Intelligence (SETI) project has been monitoring radio signals for decades. Perhaps the best explanation for the "great silence" is that, improbably or not, we really are alone.

The argument is similar to one that is often advanced against time travel. If time travel is possible then eventually its technology will be discovered. But there have been no visitors from that future. A reasonable explanation is because it's impossible. There are, of course, alternative explanations for our failure to observe visitors from other times and other planets. Maybe they're already here but concealing themselves for some reason, or maybe they know about us but consider us unworthy of their acquaintance. But these explanations are fanciful and completely ad hoc. A better alternative to the hypothesis that time travel is impossible is that our civilization will end before it achieves the technology or ingenuity for backwards time travel. A parallel explanation could be offered for the absence of extraterrestrials: either intelligent life is for some reason extremely unlikely to arise in the first place or the advanced civilizations that might have overlapped ours were all destroyed before they developed interstellar communication. Either way, we have additional reason to take the risks to humanity seriously: either intelligent life is exceedingly rare or advanced civilizations are very delicate.

FINE-TUNING, THE MULTIVERSE, AND PANSPERMIA

As we saw in chapter two, proponents of Intelligent Design argue that certain biological structures are too functionally complex to have occurred by chance. Some cosmologists employ similar reasoning, though not in support of an intelligent designer. Rather they argue that the physical and biological conditions that were necessary for the existence of life and consciousness are so improbable in themselves that this must be only one of many universes. If the fundamental forces had been even slightly weaker or stronger, or had the rate of expansion of the universe shortly after the big bang been slightly different, then there would not have existed the elements necessary for life: carbon, oxygen, hydrogen, and so on. The basic idea of the multiverse hypothesis is that although each of the physical constants is improbable in themselves, none of them is unexpected or surprising since they all obtain. Suppose your boss has won the annual office raffle three years running. You might begin to suspect that it's rigged – until you find out she always buys nearly all the tickets for sale. It's not surprising that the winning ticket is hers because they're nearly all hers. Similarly, we should not be surprised that this universe exists because they all do. Multiverse theory can be supplemented by anthropic reasoning to explain why we occupy one of the rare life-permitting universes: what other kind of universe could we observe?

Similar puzzles arise in the biological context about the origin of life: it is exceedingly unlikely that the inorganic molecules should randomly combine to form amino acids, proteins, and nucleic acids (such as DNA and RNA). According to the calculations of the groundbreaking astrophysicist Fred Hoyle, the chance of amino acids forming a common protein is comparable to the chance of a galaxy full of blind men simultaneously solving the Rubik's Cube puzzle. Hoyle's estimations have been questioned, of course, and the origin of life remains an active area of inquiry in molecular biology. Hoyle, however, was so impressed by the

FINE-TUNING, THE MULTIVERSE, AND PANSPERMIA (*cont.*)

numbers that he proposed the hypothesis of "panspermia," that the elements of life are naturally ubiquitous in the universe and that life on earth originated from outer space. The idea is controversial, although it was endorsed by Francis Crick, co-discoverer of DNA. So is the multiverse hypothesis. Both hypotheses aim to show that otherwise highly improbable (but fortuitous) circumstances are unsurprising consequences of unobserved background phenomena.

There are alternative explanations for fine-tuning, such as the theist's design hypothesis. Another possibility is that the improbability of our universe does not need to be explained at all. Given we exist, there is a sense in which it cannot be surprising that the universe made this possible, as friends of the Anthropic Principle have urged. Or perhaps it really is just an extremely improbable "brute fact" that our universe allows life and intelligence. After all, life itself is full of unlikely events: any deal of cards, the collection of events that brought your parents together, your own genetic profile. Perhaps the fact that our universe permits life is nothing more than a cosmic fluke.

Transcending humanity

It is undeniable that human existence has been transformed, and in many ways improved, by science and technology. Those living in developed economies have longer, healthier, and more comfortable lives, with greater opportunities for travel, intellectual and cultural exploration, and simple entertainment. These improvements have so far come mostly from the transformation of the human environment by technologies derived from science: safer and more abundant food and water, climate-controlled shelter, medicines to stave off dangerous viruses and bacteria, and dazzling electronic diversions. But increasingly science is exploring more "direct" means of improving human existence: improving humans themselves. Genetic manipulation of human embryos in vitro promises not only therapeutic prevention of genetic diseases but also "enhancement" of desirable traits like height and general intelligence. Family planning of the future may rely on a mix of genetic manipulation, intentional importing of "quality" genes from sperm and egg markets, cloning, and good old-fashioned sexual reproduction. We will no doubt be eager to improve the genetic prospects of our children and grandchildren, just as we might now invest in their financial futures.

There are other science-driven opportunities for human enhancement besides gene therapy. "Cosmetic neurology" via pharmaceuticals that improve memory and cognitive skills may soon be available even for those suffering no brain impairment or damage. The crude forms of steroids, hormones and stimulants now employed by athletes will give way to safer and more effective drugs for improved physical, cognitive (and sexual) performance. Safe and non-addictive manufactured opiates and narcotics, as well as antidepressants, may someday replace alcohol and nicotine as the cocktails of choice. Internal and external methods of stimulating the brain electrically, currently under investigation for Alzheimer's and epilepsy sufferers, could be extended to relieve run-of-the-mill "brain fog." And various modes of "brain-machine interface" (BMI) such as the cochlear implants commonly used to treat hearing loss may allow humans to vastly sharpen and extend sense perception and motor control.

The standard objection to human enhancement is that it undermines the inherent dignity of humans by implying that, like artifacts, some are in need of improvement. The difficulty with this objection is unpacking the notion of human dignity in such a way that it permits treatment of genetic and other defects while prohibiting enhancement and cosmetics. Another objection to enhancement technologies, which applies to advanced medical technologies generally, is the concern that human enhancement will serve primarily the most economically advantaged. A special worry about genetic enhancement is that it could allow unequal improvement of the already advantaged ethnic and social groups, and thus amount to a kind of "liberal eugenics" as one defender of enhancement has put it.

As I mentioned at the beginning of this chapter, some researchers speculate that the increased intelligence brought by the first generation of human enhancement technologies will trigger a rapid acceleration of enhancement: the smarter we become, the better we get at becoming smarter. Humans may one day "transcend" humanity in the sense that cognition becomes largely dependent on non-organic technology, possibly spread through networks like the internet, and reproduction is taken over by direct genetic engineering no longer subject to evolutionary constraints. Where will all this enhancement lead? The leading philosophical "transhumanist" (sometimes "posthumanism" is the preferred label) Nick Bostrom imagines a "Letter from Utopia" advertising the joys of our transhuman future:

My consciousness is wide and deep, my life long. I have read all your authors – and much more. I have experienced life in many forms and from many angles: jungle and desert, gutter and palace, heath and suburban creek and city back alley. I have sailed on the high seas of cultures, and swum, and dived ... You could say I am happy, that I feel good. You could say that I feel surpassing bliss. But these are words invented to describe human experience. What I feel is as far beyond human feelings as my thoughts are beyond human thoughts. I wish I could show you what I have in mind. If only I could share one second of my conscious life with you!

Sounds pretty good. There are risks of course that accompany transhuman progress, dystopian scenarios familiar to readers of science fiction. The technology might massively fail, or fall into the wrong hands, or bring down the rages of a jealous god lying in wait. But even if all goes well it is far from clear that the transhumanist utopia is a place we should look forward to. In a more analytical discussion of the future of humanity, Bostrom identifies a number of features of "the posthuman condition," including:

- Complete control of sensory input for the majority of people for most of the time.
- Human psychological suffering becomes a rare occurrence.
- Life expectancy greater than 500 years.

Consider the first of these conditions. Through the use of advanced BMI, I could produce whatever experience I wanted: world travel, scientific research, or a hot date. And since my awareness that the experience is programmed might cheapen it, or bring cruel disappointment upon its completion, I may choose to have a long-term simulation programmed that is indistinguishable from ordinary life, except much more interesting and satisfying. Bostrom maintains that such simulation is a likely destination for posthuman existence. (Indeed, he argues we may already even be posthumans living in a massive simulation designed for nostalgia or historical research!)

Would we really choose to have our experience produced in this way, mediated and enhanced through BMI, or entirely constructed from whole cloth through simulation? The question was examined carefully by the philosopher Robert Nozick thirty years ago. Nozick was concerned with the nature of ultimate value. To raise doubts about the hedonistic view that pleasure is the highest good, he asked whether we would want to spend part or all of our lives in an "experience machine" (EM). We would float in a tank, with our brains attached to a computer programmed to give us whatever experience we desired. From the inside, we would have lives of great pleasure, excitement and fulfillment. From the outside, we would be pale, wrinkled and atrophied, kept alive by an IV drip and life-support machines.

Confronted with this hypothetical offer, most say they'd prefer strongly to remain in the real world despite the imperfections and disappointments. To be sure, some standard reasons for hesitating to enter the experience machine do not implicate the posthuman condition: leaving behind friends and family, the letdown when we exit, the potential incompetence or malevolence of EM technicians, etc. But there is a deeper reason for staying outside: we want not merely to *seem* to travel, help others, produce works of art; we want to *really do* these things. This seems especially clear for compassionate service and learning, two activities whose value seems to be constituted not merely by the experience that accompanies them but also by their relations to their objects. If Mother Theresa or Albert Einstein awoke from a pleasant dream and realized their greatest achievements had nothing to do with the real world, one doubts they would look back fondly on their reverie. So why would we look forward to such experiences? By putting sensory inputs under our control, and possibly delivering us altogether from our bodies, the posthuman condition deprives us of the causal connections to the world that make certain experiences supremely important.

Even granting the admittedly elusive notion that connection with the "real" is valuable over and above all experiences, the posthumanist will object that the distinction between mere experience and reality is just a continuum. Even in the waking life of ordinary humans, our relations to the "real world" are highly mediated and to some degree within our control. Our perceptions are structured by our language, concepts, and theories and also colored by our upbringing and personality.

Humans have always controlled their sensory inputs using art, travel, intoxicants, and so on. What future technology offers is merely a more effective set of tools for fashioning happier lives from the world we've been given.

But what about the second feature of the posthuman condition: should it be an aim of science that "human psychological suffering becomes a rare occurrence"? If the chronic depression, fear and self-loathing that plague so many modern humans will disappear in the posthuman era that's cause for celebration. But it isn't so clear that we should attempt to make psychological suffering "rare." Along with the familiar saw about "suffering for art," one might add that for most humans there is a seemingly inextricable connection between various kinds of psychological suffering (disappointment, error, regret, anxiety) and important goods (achievement, wisdom, nostalgia, relief). It is possible, of course, that future psychology will make feelings of achievement not depend upon prior strivings and failures. But then we face the previous worry that a "manufactured" sense of well-being is not something to aim for, not our collective *summum bonum*.

The third condition of posthumans, living 500+ years or perhaps forever, is tantalizing but also problematic. Proponents of transhumanism note that the promise of immortality has long been a stronghold (and cash-cow) for religions, and it does seem that on the whole people would rather die later than sooner. But do we really want to live forever, even freed from the all-too-familiar effects of physical and mental aging? Certainly we would have occasion to pursue literature, art and science to an extent that would otherwise have been impossible. But having gone through all the major works of literature and music, over and over again, mastered science and mathematics, and explored *ad nauseum* all the varieties of human relationships, wouldn't life become tedious and burdensome? It is true that one might develop new, unexpected interests after 500 years of life. But do interests I might develop in the distant posthuman future, interests I do not have at present, really give *me* a reason to want to live then? As the philosopher Bernard Williams has emphasized, when we contemplate a very long life we face an unavoidable dilemma: either we will exhaust all our interests and become intolerably bored; or in the distant future we will have interests and concerns so radically different that it is not so much *my* future as the future of a different person who evolves *from* me.

Whether or not we are destined for posthumanity, nothing lasts forever. Since the nineteenth century physicists have speculated that the ultimate fate of the universe is a "heat death." As existing sources of energy burn out, heat slowly spreads evenly through the universe in accordance with the second law of thermodynamics. Whether the universe goes on expanding forever, as seems likely, or collapses with a "big crunch," organic life will become impossible in the relative short run as the stars burn out in a trillion years or so and even atoms disintegrate. Even if humans are "uploaded" into robots or computers, no organized information processing will be possible as heat death approaches (although some have speculated about a "subjective" eternity for certain kinds of intelligent systems).

The seeming inevitability of heat death is, as Bostrom has put it, "a matter of some personal concern" to the posthumanists. A very different attitude toward death is expressed by the ancient Greek philosopher Epicurus: "Death is nothing to us. Where we are death is not; where death is we are no longer." Epicurus' point is that fear of death comes from imagining it to be a future state, perhaps a very boring or distressing one. But death is not a state that I'll be in – it's the absence of an "I" to have states. My not existing shouldn't bother me since it can't possibly happen to me. Epicurus' follower the Roman philosopher Lucretius offered a therapeutic thought experiment for those who are anxious about death: think about the centuries of time that passed before you were born. No one curses their non-existence before they were born; but this is a perfect mirror of death.

Perhaps what depresses people about the prospect of human extinction is not their own personal demise but the final destruction of human culture: no more great music, literature, or science. In a few billion years it will be as if none of that ever existed. True. But the fact that things will not matter in the future doesn't seem to make them matter any less now. Lucretius might point out that art and science did not exist billions of years ago. Maybe it would have been good if they had, but we certainly don't consider it a tragedy that they didn't.

Perhaps there is a crucial difference between the past and future absence of humanity: extinction is approaching but the pre-human era is behind us. However, the conception of time implicit in this difference – that it flows from the past into the future – however much ingrained in human perception, isn't really countenanced in modern physics. If we consider the universe as an enormous four-dimensional space-time manifold, we do not find a privileged "now" that is moving from earlier times to later ones. At what rate would it flow, after all? Considered objectively, or "from the point of view of eternity," as the philosopher Spinoza says, all times are equally real, including the long times before and after the existence of humans and the brief period of human flourishing. From this point of view, we can offer ourselves the consolation that Einstein offered to the newly widowed wife of his friend Michele Besso: "Now he has departed from this strange world a little ahead of me. That means nothing. People like us, who believe in physics, know that the distinction between past, present, and future is only a stubbornly persistent illusion." When life ceases human culture still exists, just at an earlier time. To wish that it will exist at all times is asking too much, like wishing that always and everywhere it is Paris in the springtime.

Further reading

1 Origins of science

Ancient science

Quite readable, high quality histories of Ancient science include *Greek Science in Antiquity,* by Marshall Clagett (Abelard-Schuman Inc, 1955); *Early Greek Science: Thales to Aristotle* by G. E. R. Lloyd (W.W. Norton, 1970); and *The Beginnings of Western Science,* by David C. Linderg (University of Chicago, 1992). For more technical and scholarly studies, see: *Ancient Science through the Golden Age of Greece,* by George Sarton (Dover Publications, 1993); and *The Exact Sciences in Antiquity,* by Otto Neugebauer (Brown University Press, 1957).

Medieval and Renaissance science

The following provide reliable accounts of medieval science through the Renaissance: *A History of Natural Philosophy,* by Edward Grant (Cambridge University Press, 2007); *The Science of Mechanics in the Middle Ages,* by Marshall Clagett (University of Wisconsin Press, 1959); *Causality and Scientific Explanation,* Vol. 1: *Medieval and Early Classical Science,* by William A. Wallace (University of Michigan Press, 1972); *The Foundations of Modern Science in the Middle Ages,* by Edward Grant (Cambridge University Press, 1996); *The Scientific Renaissance, 1450–1630,* by Marie Boas Hall (Dover Publications, 1962); and *Man and Nature in the Renaissance,* by Allen Debus (Cambridge University Press, 1978). A one-volume abridgement and translation of Pierre Duhem's magisterial eleven-volume history of cosmology from Plato to Copernicus, *Système du Monde* is now available: *Medieval Cosmology*, edited and translated by Roger Ariew (University of Chicago Press, 1985).

The Copernican Revolution

Galileo's *Dialogues Concerning the Two Chief World Systems* is one of the most intellectually exciting documents in the history of science. The standard English translation is by Stillman Drake (Modern Library, 2001). An excellent collection of primary documents, including the most important "letters," is *The Galileo Affair: A Documentary History,* edited by Finocchiaro (University of California Press, 1989). The modern cosmologist Stephen Hawking has collected a number of the more technical treatises of the Copernican Revolution in *On the Shoulders of Giants: The Great Works of Physics and Astronomy* (Running Press, 2003). There are many fine, accessible histories of the Copernican Revolution: *Sleepwalkers*, by Arthur Koestler (Penguin, 1990); *From the Closed World to the Infinite Universe*, by Alexandre Koyré (Johns Hopkins, 1968); and *The Copernican Revolution*, by Thomas Kuhn (Harvard, 1992). *Galileo at Work*, by Stillman Drake (Dover, 2003) is a fine scientific biography; *Galileo, Bellarmine and the Bible*, by Richard J. Blackwell (Notre Dame Press, 1992) explains the religious context of the Galileo affair; *Galileo's Daughter*, by Dava Sobel is a moving and scholarly portrait (Penguin, 2000); *Galileo, Courtier*, by Mario Biagioli (Chicago, 1994) presents a controversial, rather unflattering account of Galileo's career.

The Scientific Revolution

Descartes' *Discourse on Method* (Hackett Publishing, 1999) provides a taste of his philosophy and his science, as, respectively, do William Harvey's *Circulation of the Blood* (Elsevier,

1971) and Robert Boyle's *Philosophical Papers* (Hackett, 1991). The standard English edition of Newton's *Principia*, not an easy read, is edited by I. B. Cohen and Anne Whitman (University of California, 1999); Andrew Janiak provides a useful collection of Newton's *Philosophical Writings* (Cambridge, 2004); a recent edition of the *Leibniz-Clarke Correspondence* is provided by Roger Ariew (Hackett, 1999). An exhaustive Newton biography is *Never at Rest*, by Richard Westfall (Cambridge, 1983); James Gleick's *Isaac Newton* (Vintage, 2003) is much briefer. Here are some well-known overviews of the Scientific Revolution: *The Metaphysical Foundations of Modern Science*, by E. A. Burtt (Dover, 2003); *The Construction of Modern Science*, by Richard Westfall (Cambridge, 1978); *The Revolution in Science* by A. Rupert Hall (Addison Publishing, 1983); *The Scientific Revolution*, by Steven Shapin (University of Chicago, 1996); *The Mechanization of the World Picture*, by E. J. Dijksterhius (Princeton, 1986); *The Bible, Protestantism and the Rise of Natural Science*, by Peter Harrison (Cambridge, 2001).

2 Defining science

Testability and demarcation

Karl Popper: *The Logic of Scientific Discovery* (Routledge, 2002); *Conjectures and Refutations* (Routledge 2002); *The Open Society and Its Enemies* (Routledge, 2006). For a detailed explication and defense of Popper's theory of science, see *Critical Rationalism*, by David Miller (Open Court, 1994). Thomas Kuhn: *Structure of Scientific Revolutions* (University of Chicago, 1996); Lakatos: *Methdology of Scientific Research Programmes* (Cambridge University, 1978); *Freud and the Question of Pseudoscience*, by Frank Cioffi (Open Court, 1999); *Why People Believe Weird Things*, by Michael Shermer (Holt, 2002).

Intelligent design

On the classical design argument, see the defense by William Paley, *Natural Theology* (Oxford University Press, 2008), and the critique by David Hume, *Dialogues Concerning Natural Religion* (Hackett, 1998). The main defenses of ID are *Darwin on Trial*, second edition, by Phillip Johnson (InterVarsity Press, 1993), *Darwin's Black Box*, second edition, by Michael Behe (Free Press, 2006) and *The Design Inference*, by William Dembski (Cambridge University Press, 2006). For philosophical criticism of ID, see *God, the Devil and Darwin,* by Niall Shanks (Oxford University Press, 2003); *Living with Darwin*, by Philip Kitcher (Oxford University Press, 2007) and *Darwinism and Its Discontents*, by Michael Ruse (Cambridge University Press, 2008).

String theory

Detailed criticisms of string theory coming from physicists include *The Trouble with Physics*, by Lee Smolin (Mariner Books, 2006) and *Not Even Wrong*, by Peter Woit (Basic Books, 2006). Popular and informed defenses of string theory, which directly engage the testability question, are *The Elegant Universe*, by Brian Greene (Norton, 2003) and *Warped Passages*, by Lisa Randall (Harper Collins, 2005).

3 Scientific method

Deductivism vs. inductivism

For primary works by Newton and Descartes see Further reading, chapter one. For good discussions of their methods see *The Newtonian Revolution*, by I. B. Cohen (Cambridge University Press, 1983) and *Descartes' Philosophy of Science*, by Desmond Clarke (Manchester University Press, 1982).

Bacon's inductive machine

Bacon's most important work on method is the *New Organon*, edited by Lisa Jardine and Michael Silverthorne (Cambridge University Press, 2000). For a concise overview of Bacon's method, which argues that he is not entirely opposed to hypotheses, see *Francis Bacon's Philosophy of Science*, Peter Urbach (Open Court, 1987).

Mill's methods

Mill's *System of Logic* comprises volumes 7 and 8 of the *Collected Works*, edited by J. M. Robson and R. F. McRae (University of Toronto Press, 1973). Almost any good logic text will provide an introduction to his methods of causal reasoning, for example, *A Concise Introduction to Logic*, ninth edition, by Patrick Hurley (Wadsworth, 2005). For a summary of Mill's Inductivism see the article by Geoffrey Scarre in the *Cambridge Companion to Mill*, edited by John Skorupski (Cambridge University Press, 1998). Whewell's major writings on method, including his critique of Mill, are collected in *Theory of Scientific Method*, edited by Robert E. Butts (Hackett, 1968).

Popper's deductivism

Along with the works of Popper listed in Further reading, chapter two, the reader may wish to look at the broad collection of brief excerpts *Popper Selections*, edited by Popper's student David Miler (Princeton, 1985). For a recent "critical assessment" of Popper's philosophy, see *Karl Popper*, by Anthony O'Hear (Routledge, 2003).

Hempel's hypothetico-deductivism

Several of Hempel's major papers on method were reprinted in *Aspects of Scientific Explanation* (Free Press, 1965). Hempel's treatment of some standard problems in (mid-twentieth century) philosophy of science can be found in *Philosophy of Natural Science* (Prentice-Hall, 1966). For a collection of scholarly papers on the history of logical empiricism, see *Cambridge Companion to Logical Empiricism*, edited by Alan Richardson and Thomas Uebel (Cambridge University Press, 2007).

Relativism and anarchism

Along with the classic *Structure of Scientific Revolutions* (University of Chicago Press, third edition, 1996), there is also a collection of important papers, *The Essential Tension* (University of Chicago, 1977). There are numerous studies available of Kuhn's work and influence, for example *Thomas Kuhn*, by Alexander Bird (Princeton, 2000). A good recent collection of papers on Kuhn is *Thomas Kuhn*, edited by Thomas Nickles (Cambridge University Press, 2003). N. R. Hanson's most influential work is *Patterns of Discovery* (Cambridge University Press, 1958). Along with the major work *Against Method*, revised edition (Verso Press, 1988), Feyerabend's autobiography *Killing Time* (University of Chicago Press, 1996) is a very good read.

Holism and naturalism

For some of Duhem's most important writings in philosophy of science, see Pierre Duhem: *Essays in the History and Philosophy of Science*, edited by Roger Ariew and Peter Barker (Hackett, 1996). Quine's holism and naturalism are developed in papers contained in *From a Logical Point of View* (Harvard University Press, 1953) and *Ontological Relativity* (Columbia University Press, 1969). Ronald Giere's "naturalized philosophy of science" is defended in his *Explaining Science* (University of Chicago, 1988). For a collection of papers in "experimental philosophy" see *Experimental Philosophy*, edited by Joshua Knobe and Sean Nichols (Oxford University Press, 2008).

4 The aims of science

Scientific realism

A good recent discussion and defense of scientific realism is *Scientific Realism: How Science Tracks Truth*, by Stathis Psillos (Routledge, 1999). A collection of influential papers for and against realism is *Scientific Realism*, edited by Jarret Leplin (University of California Press, 1984). Arthur Fine's charge that the no miracles argument begs the question is made in "Piecemeal Realism," *Philosophical Studies* 61: 79–96 (1991) and in "The Natural Ontological Attitude," his contribution to the Leplin volume. DeMorgan explains the "opponent fallacy" in *Formal Logic* (Taylor and Walton, 1847). The classic statement of the pessimistic induction is by Larry Laudan, "A Confutation of Convergent Realism" in Leplin's collection. Versions of the underdetermination argument besides Mill's (see chapter three) can be found in Pierre Duhem, *The Aim and Structure of Physical Theory* (Princeton University Press, 1991) and Bas Van Fraassen, *The Scientific Image* (Clarendon Press, 1980), which also contains Van Fraassen's evolutionary explanation for the success of science. *Images of Empiricism*, edited by Bradley Monton (Oxford University Press, 2006) offers recent reflections on constructive empiricism.

For an interesting recent defense of anti-realism based on the possibility of "unconceived alternatives," see *Exceeding our Grasp*, by P. Kyle Stanford (Oxford University Press, 2006). A good recent edition of Descartes' *Meditations* is by John Cottingham et al. (Cambridge University Press, 1996). Versions of progress-realism are defended by Popper, for example *Objective Knowledge* (Oxford University Press, 1972), Niinliluoto, *Truthlikeness* (D. Reidel, 1987) and Kitcher, *The Advancement of Science* (Oxford, 1993). For an influential defense of structural realism, see Worrall, "Structural realism: The best of both worlds?" *Dialectica* 43: 99–124 (1989).

Varieties of anti-realism

For classic versions of instrumentalism, see Duhem, *The Aim and Structure of Physical Theory* and E. Mach, *The Science of Mechanics* (Open Court, 1960). Good discussions of reductionism of various kinds may be found in Hempel, *Aspects of Scientific Explanation* (see Further reading, chapter two), Nagel, *The Structure of Science* (Routledge, 1961) and Suppe (ed.), *The Structure of Scientific Theories* (University of Illinois Press, 1977). Van Fraassen formulates and defends constructive empiricism in *The Scientific Image*. See also *The Empirical Stance* (Yale University Press, 2004). *Images of Science*, edited by Paul Churchland and Clifford Hooker (University of Chicago, 1985) is a good collection of critical commentaries on Van Fraassen's view. Kuhn's argument for conceptual relativism is presented in the *Structure of Scientific Revolutions* (see Further reading, chapter two). For some more recent reflections of Kuhn, in which he moderates his views somewhat, see *The Road Since Structure* (University of Chicago, 2002). Goodman's version of relativism is presented in *Ways of Worldmaking* (Harvard University Press, 1978). For a defense of unification as a form of explanation, see Kitcher, "Explanatory unification and the causal structure of the world" in P. Kitcher & W. Salmon, eds, *Scientific Explanation* (University of Minnesota Press, 1989). See also the texts by Nagel and Suppe mentioned above. John Dupré, *The Disorder of Things* (Harvard University Press, 1993) provides a critical appraisal of reductionism and unity of science. For an overview of behaviorism, see Skinner's *Science and Human Behavior* (Macmillan, 1953). Chomsky provided an influential critique of behaviorism in his "Review of Verbal Behavior," *Language*, 35, 26–58 (1959). Fodor's criticisms of reduction and unity are given in "Special sciences, or the disunity of science as a working hypothesis," *Synthese* 28: 77–115 (1974). For a straightforward defense of emergence, see *Mind and the Emergence*, by

Philip Clayton (Oxford University Press, 2006). For a variety of pluralist perspectives on the aims of science, see *Scientific Pluralism*, edited by Stephen Kellert, Helen Longino and Kenneth Waters (University of Minnesota Press, 2006).

5 Social dimensions of science

Sociology of science

The Sociology of Science, by Robert Merton (University of Chicago, 1979). Along with the *Structure of Scientific Revolutions*, Kuhn discusses social aspects of science in "Objectivity, Values and Theory Choice" in *The Essential Tension* (University of Chicago, 1977). For an (opinionated) overview of the relation between philosophy of science and sociology of science, see *Social Epistemology*, second edition, by Steve Fuller (Indiana University Press, 2002).

The strong programme

Knowledge and Social Imagery, second edition, by David Bloor (University of Chicago, 1991); *Scientific Knowledge: A Sociological Approach*, by Barry Barnes, David Bloor and John Henry (University of Chicago, 1996); *Leviathan and the Air Pump*, by Steven Shapin and Simon Schaeffer (Princeton University Press, 1989); *A Social History of Truth*, by Steve Shapin (University of Chicago, 1995).

Social constructivism

Laboratory Life, by Bruno Latour and Steve Woolgar (Princeton University Press, 1986); *Pandora's Hope: Essays on the Reality of Science Studies*, by Bruno Latour (Harvard University Press, 1999); "Has Critique Run out of Steam?," by Bruno Latour,

Social Inquiry 30: 225 – 248 (2004); *The Social Construction of What?* by Ian Hacking (Harvard University Press, 1999); *Social Constructivism and the Philosophy of Science*, by André Kukla (Routledge, 2000). *Defending Science – Within Reason*, by Susan Haack (Prometheus, 2003).

The Sokal hoax

Sokal's original article, his exposé, a response from the *Social Text* editors, and commentaries including by Stanley Fish, Phillip Kitcher, and Steven Weinberg may be found in *The Sokal Hoax*, edited by *Lingua Franca* (Brison Press, 2000); Sokal's most recent book is *Beyond the Hoax: Science, Philosophy and Culture* (Oxford, 2008). The book that inspired Sokal is *Higher Superstition: The Academic Left and Its Quarrels with Science*, by Paul Gross and Norman Levitt (Johns Hopkins University Press, 1994).

Feminist philosophy of science

Science and gender

On the association of men with reason and mind, and women with emotion and body, see *The Man of Reason*, by Genevieve Lloyd (University of Minnesota Press, 1984); *Death of Nature*, by Carolyn Merchant (Harper and Row, 1980); *Science as Social Knowledge*, by Helen Longino (Princeton University Press, 1990). See also *The Fate of Knowledge* (Princeton, 2002); *Science and Gender*, by Ruth Bleier (Pergamon, 1984); *Primate Visions*, by Donna Haraway (Routledge, 1989); *Thinking From Things*, by Alison Wylie (University of California, 2002); *The Woman that Never Evolved*, by Sarah Hrdy (Harvard University Press, 1981); *Gender and Boyle's Law of Gases*, by Elizabeth Potter (Indiana University Press, 2001); *A Feeling for the Organism: The*

Life and Work of Barbara McClintock, by Evelyn Fox Keller (Freeman, 1983).

What is feminist science?

Books by Helen Longino are cited in the section above. *Reflections on Science and Gender*, by Evelyn Fox Keller (Yale University Press, 1985); Sanrda Harding, *The Science Question in Feminism* (Cornell University Press, 1986). See also *Whose Science, Whose Knowledge?* (Cornell University Press, 1991). For a recent overview see *Feminism and Philosophy of Science*, by Elizabeth Potter (Routledge, 2006). See also *Illusions of Paradox*, by Richmond Campbell (Rowan & Littlefield, 1998).

Science and values

The fact value distinction

Treatise of Human Nature, by David Hume (Oxford University Press, 2000); *Principia Ethica*, by G. E. Moore (Dover, 2004); *Why I am not A Christian and Other Essays,* by Bertrand Russell (Barlow Press, 2008); *The Unity of Science*, by Rudolph Carnap (Kegan Paul, 1934); *Collapse of the Fact/Value Dichotomy*, by Hilary Putnam (Harvard University Press, 2004).

The impact of values on science

Catastrophe, by Richard Posner (Oxford University Press, 2004). LHC scientists defend its safety in the report "Review of the Safety of LHC Collisions," by the LHC Safety Assessment Group (John Ellis et al.), available at: lhc.web.cern.ch/lhc/. For good philosophical analysis of "IQ" see *The IQ Controversy*, edited by Ned Block and Gerald Dworkin (Pantheon 1976). *The Bell Curve*, by Richard Herrnstein and Charles Murray (Simon &

Schuster, 1996) suggested that IQ difference was a factor in economic inequality among races in the U.S. The well-known biologist Stephen Jay Gould offers a critique of *The Bell Curve* and generally of the usefulness of the concept of "general intelligence" in biology and social science in *The Mismeasure of Man* (W. W. Norton, 1996). *The Lysenko Affair*, by David Joravsky (University of Chicago Press, 1970). On the Bush Administration's supposed intrusion upon science, see *The Republican War on Science* (Basic Books, 2005). For a recent collection of philosophical papers mostly defending the role of non-epistemic values in science, including the paper by Wylie and Nelson, see *Value-Free Science?*, edited by H. Kincaid, J. Dupré, and A. Wylie (Oxford University Press, 2007).

The impact of science on values

Spinoza, *Ethics* (Penguin Classics, 2005); *The Criminal Prosecution and Punishment of Animals,* by E. P. Evans (Lawbook Exchange, 1998); *Bioethics and the Brain,* by W. Glannon (Oxford University Press, 2006); *The Ethical Brain,* by M. S. Gazzinga (Dana Press, 2005); "The Brain on the Stand," by Jeffrey Rosen, *New York Times,* March 11, 2007.

6 Science and human futures

Are we doomed?

Nick Bostrom offers his prognosis in "The Future of Humanity," in *New Waves in Philosophy of Technology*, edited by Jan-Kyrre Berg Olsen and Evan Selinger (Macmillan, 2008). See also *Global Catastrophic Risks,* edited by Nick Bostrom and Milan Cirkovic (Oxford University Press, 2008). On the notion of runaway technological progress, or so-called "singularity," see *The Singularity is Near*, by Ray Kurzweil (Viking, 2005). For recent overviews of catastrophic risks, see *The End of the World*, by John Leslie (Routledge, 1996); *Our Final Hour*, by Martin Rees (Basic Books, 2003).

Expected costs

Risk and Rationality, by Katherine Shrader-Frechette (University of California Press, 1991). *Catastrophe*, by Richard Posner (Oxford University Press, 2004) looks at massive threats in terms of risk, policy, and the law. See also *Blindside*, edited by Francis Fukayama (Brookings Institution Press, 2007). On the precautionary principle, *Laws of Fear: Beyond the Precautionary Principle*, by Cass Sunstein (Cambridge University Press, 2005) and T. O'Riordan and J. Cameron, *Interpreting the Precautionary Principle* (Earthscan Publications, 1995). For a very readable critique of standard cost-benefits analysis in environmental policy, and a defense of the precautionary principle, see *Priceless*, by Frank Ackerman and Lisa Heinzerling (New Press, 2004).

Future generations

There is a massive philosophical literature on obligations to future generations. For a good sampling of classic papers, see *Responsibilities to Future Generations*, edited by E. Partridge (Prometheus, 1981), and *Obligations to Future Generations*, edited by R. I. Sikora and B. Barry (Temple University Press, 1978). Derek Parfit presents an influential analysis of this and related moral problems in *Reasons and Persons* (Clarendon Press, 1984). A recent work focused on social policy is *Environmental Justice and the Rights of Unborn and Future Generations*, by Laura Westra (Earthscan, 2006).

The tragedy of the commons

Classical discussions of the problem of justifying cooperation include, Plato, *The Republic*, Bk II, translated by C. D. C. Reeve

(Hackett, 2004) and Thomas Hobbes, *Leviathan* (Oxford University Press, 1998). The touchstone modern formulation of the problem is by Garrett Hardin, "The Tragedy of the Commons," *Science* 162: 1243–1248 (1968). See also *Collective Action*, by Russell Hardin (Johns Hopkins Press, 1982); *Evolution of the Social Contract*, by Brian Skyrms (Cambridge University Press, 1996); *Morals by Agreement*, by David Gauthier (Clarendon Press, 1986).

Evolution of cooperation

The cheaters/suckers/grudgers example is used by Dawkins in *The Selfish Gene* (Oxford University Press, 1976) and developed by John Mackie, "The Law of the Jungle," in *Philosophy* 53: 455–464 (1978). The concept of "reciprocal altruism" was developed by Robert Trivers. See, for example, "The evolution of reciprocal altruism," *Quarterly Review of Biology* 46: 35–57 (1971); Robert Axelrod's pioneering work in game theory, *The Evolution of Cooperation* (Basic Books, 1984) explains iterated prisoners' dilemmas and the "tit for tat" strategy. For an interesting recent philosophical discussion of the evolution of morality see Richard Joyce, *The Evolution of Morality* (M.I.T. 2006).

Colonization of space

Frank Tipler defends interstellar travel using nanobots and "Von Neumann probes" containing our genetic or neurophysiological profiles in *The Physics of Immortality* (Anchor, 1997). See also Freeman Dyson, *Infinite in All Directions* (Harper and Row, 1988); Carl Sagan, *Pale Blue Dot* (Random House, 1994); *Our Cosmic Future*, by Nikos Prantzos (Cambridge University Press, 1998). On the puzzle of whether our "personal identity" could be retained in scenarios relevantly similar to those discussed by Tipler, see Derek Parfit, *Reasons and Persons*.

Are we alone?

Paul Davies discusses the argument for infinite duplicates, which he attributes to the physicist G. F. R. Ellis, in *Are We Alone?* (Basic Books, 1995). See also "Philosophical Implications of Inflationary Cosmology," Joshua Knobe, Ken D. Olum and Alexander Velenkin, *British Journal for the Philosophy of Science*, 57: 47–67. For a good collection of scientific papers on both extraterrestrial life and space travel, see *Extraterrestrials: Where Are They?*, edited by Ben Zuckerman and Michael Hart (Cambridge University Press, 1995). For a discussion emphasizing the role of the SETI project, see *Is Anyone Out There?*, by Frank Drake and Dava Sobel (Delacorte Press, 1992). Brandon Carter discusses the anthropic principle in "Large Number Coincidences and the Anthropic Principle in Cosmology," in *Physical Cosmology and Philosophy*, edited by John Leslie (Macmillan, 1990). For a detailed philosophical discussion of anthropic reasoning, see *Anthropic Bias*, by Nick Bostrom (Routledge, 2002). On Hoyle's panspermia account of the origins of life, see *The Intelligent Universe* (Holt, Reinhart and Winston, 1984).

Enhancing and transcending humanity

Two good philosophical assessments of enhancement, especially genetic enhancement, are *The Case against Perfection*, by Michael Sandel (Harvard University Press, 2007), and *Choosing Children*, by Jonathan Glover (Oxford University Press, 2006). See also *Our Posthuman Future*, by Francis Fukayama (Farrar, Strauss and Giroux, 2002). See Leon Kass, *Life, Liberty and the Defense of Dignity* (Encounter, 2004), for a critique of enhancement. *The Hedonistic Imperative,* by David Pearce (BLTC, 2007) is an enthusiastic defense of enhancement. See also *Liberal Eugenics: In Defence of Human Enhancement*, by N. Agar (Blackwell, 2004); *Radical Evolution*, by Joel Garreau (Doubleday, 2004); and *Redesigning Humans*, by Gregory Stock (Houghton Mifflin,

2002). "Letter from Utopia," by Nick Bostrom, *Studies in Ethics, Law, and Technology* 2: 1–7 (2008). See also Bostrom's "Future of Humanity" (see above in Further reading for "Are we doomed?"). Robert Nozick's original discussion of the "experience machine" is in his *Anarchy, State and Utopia* (Basic Books, 1974). Bernard Williams' critique of immortality is in a paper called "The Makropulos Case: Reflections on the Tedium of Immortality," in *Problems of the Self* (Cambridge University Press, 1973).

Index